How Crises Shaped Economic Ideas and Policies

Nicos Christodoulakis

How Crises Shaped Economic Ideas and Policies

Wiser After the Events?

Springer

Nicos Christodoulakis
Department of International & European Economics Studies
Athens University of Economics & Business (AUEB)
Athens
Greece

Translation from the Greek language edition:
"ΟΙΚΟΝΟΜΙΚΕΣ ΘΕΩΡΙΕΣ ΚΑΙ ΚΡΙΣΕΙΣ", Economic
Theories and Crises, Athens: Kritiki Publications, 2012.

ISBN 978-3-319-16870-8 ISBN 978-3-319-16871-5 (eBook)
DOI 10.1007/978-3-319-16871-5

Library of Congress Control Number: 2015938161

Springer Cham Heidelberg New York Dordrecht London

Printed on acid-free paper

Springer International Publishing AG Switzerland is part of Springer Science+Business Media (www.springer.com)

Preface

The question of whether economic theories are capable of explaining the root causes of the global crisis in 2008 and how they will be transformed to take into account previous inadequacies continues to stir academic economists, policy makers, and the public at large. When I was asked to organize a new undergraduate course on these questions, my first reaction was to go back in history and see how economic theories had responded to major economic crises in the past. Was it possible for those adhered to the prevailing theories to foresee a crisis and tackle its consequences afterwards, or the dominant views had to be discarded and replaced by new approaches? And further to inquire whether the conclusions drawn from history could be of any use in amending the prevailing theories and better understanding the current crisis.

In the course of history, the interplay between economic theories and crises is found to be a process far more influential than one might have thought. First because it reveals the limitations of the theories prevailing in each epoch to face unexpected events, and this may act as a critical warning on how intensely current thinking should be scrutinized and continuously improved upon. Second because several theories—indeed the major theories—were catalyzed by previous major crises that led the then dominant paradigm to become obsolete. If economics is studying the relationship between ends and scarce means which have alternative uses, as famously epitomized by Lionel Robbins in his *Essay on the Nature and Significance of Economic Science*, then the evolution of economic theories might be considered as a process caused by unusually large crises on which alternative interpretations are tested.

After three years of developing and teaching the subject in the Athens University of Economics and Business, I decided to write this book not just as a supporting material for the course but also as contribution to the debate on how the evolution of economics should be presented so as to be better understood. A caveat is due that the book is neither a concise history of economics nor a treatise on methodology and comparative assessment of competing theories, though it uses several of such traits. It is better to be seen as an attempt to frame past problems and interpretations into a current dialogue on how economics should respond to the current crisis and which questions need to be addressed and reexamined.

Among many books and authors that I have read in the course of years and in connection to this one, I need to note the particular benefit I took from some specific works: *The methodology of economics* by Mark Blaug provided me with the essentials of scientific criteria in examining economic theories; *The Distant Mirror* by Barbara Tuchman fascinated me in understanding how societies may turn major calamities to a process of new achievements, while I found that *The Cash Nexus* by Niall Ferguson is exemplary in explaining the interactions between events and policies over time. In writing on the evolution of economic theories, I was frequently inspired by the authoritative study by Robert Heilbroner and William Milberg on *The Making of Economic Society*, and was also influenced by the way past events are brought forward by John Kenneth Galbraith in his *History of Economics: The Past as the Present*. The motivation and intellectual guidance I have drawn from these books by far exceed the specific mentioning in the text.

The first edition of the present book appeared in Greek in 2012, and the present form has benefited from many comments and suggestions in the meanwhile. I am indebted to Professors Nicos Theocharakis, Theodore Lianos, and Antonis Liakos for useful comments and corrections on the first edition. I am also thankful to George Christidis and Irene Watson who helped me in translating and editing the book in an earlier stage and to Johannes Glaeser for his encouragement and advice for the present publication. In particular, I would like to thank Anastasia Tsadaris who patiently edited the manuscripts several times before taking the present form. Naturally, any remaining errors or omissions are entirely of my own.

Athens, Greece Nicos Christodoulakis
January 2015

Contents

Introduction: Delusions and Lessons 1

Abstract

Economic crises do not exclusively erupt in our modern era but have been occurring since early on in human societies. Though very different in their causes and effects, crises throughout history had one thing in common: they strongly influenced the way we think about economics. In fact, most economic theories sprung into life as a response to a crisis, putting aside previously held orthodoxies that failed to foretell or cope with the shock. Therefore, an account of economic thought should go in tandem with the eruption and evolution of major crises, and this is also a major challenge faced by contemporary economics in relation to the recent global crisis. To facilitate the historical examination of crises and theories, a demarcation line is set around 1500 AD to characterize the onset of modern era. Its main characteristic was that both the scale of production and the freedom of economic factors became incomparably higher than in the pre-modern age.

1.1 Crises in the Nile and the Rhine

In the summer of 2011, 17 renowned economists—all living Nobel Laureates in Economics (henceforth NLE)—gathered in a small town on the Swiss-German border to discuss the global crisis, its effects and the proper policies to combat it. A prominent invitee at the meeting the German Finance Minister Wolfgang Schauble, said that "there might well be 7 lean years ahead for the world economy", and pointed to a "trade-off between short-term pain and long-term gain".[1]

This short statement includes a sense of mystery and anxiety that have caused a lot of controversy in economic theories in the past and continues to trouble today. The first concern is whether the present recessionary cycle will indeed be succeeded

[1] At Lindau on Lake Constance that is flown through by the Rhine. *Reuters*, 28 August 2011.

N. Christodoulakis, *How Crises Shaped Economic Ideas and Policies*,
DOI 10.1007/978-3-319-16871-5_1

by a cycle of growth; otherwise some economies may not endure the pressure and collapse. The second is whether growth—if it ever returns—will be sufficiently strong to compensate for the losses inflicted by recession and what kind of policies should be implemented to maximize the extent and duration of recovery.

The mystery lies in the duration of the recession, which Mr. Schauble predicted that will last for 7 years, thus rekindling a host of superstitions surrounding that famous number shared for centuries not only by the public but frequently by philosophers, priests and mathematicians.[2]

From where did Mr. Schauble unearth number seven? There might be two hypothetical explanations, one simplistic and a more sophisticated version. Let us start with the naïve explanation: Hypothesizing that Mr. Schauble, as a key member of Christian Democrats, regularly reads the Bible, it is likely that he is influenced by the story of Pharaoh's dream on the seven lean and seven fat kine. As interpreted by Joseph, the dream was foretelling 7 years of drought and famine to be followed by 7 years of fertility and plenty. It is the oldest surviving story of economic fluctuations and, as will be examined later, it is very interesting indeed to see how it was dealt with.

If Mr. Schauble was influenced by the biblical story, he may have assumed that the economic malaise may be repeated periodically. Moreover, he may have hoped that his prediction would be easily endorsed since almost everybody in the world is familiar with the story. But, of course, he may not have thought so simplistically and instead he sought to study economic fluctuations in a more scientific way. What would he have done in this case? It is plausible that he would have chosen to carefully analyze a major representative economy over a long period of time, in order to draw some robust conclusions on the duration of crises.

Let us look, for example, at the US economy and how its GDP grew over a period of 110 years, namely from 1900 until 2009. Figure 1.1 shows that, in general, US GDP is evolving upwards, sometimes slowly, and other times faster. There are also periods of decline; in some intervals GDP falls abruptly in others by just a glimpse. As it is difficult to visually elaborate on these changes, one has to employ a technical decomposition of GDP series into a smooth trend and a cyclical deviation around it.

When actual GDP is above the trend there is a peak, when below there is a trough, i.e. a recession. Figure 1.2 displays all peaks and troughs during the 110 year period, and a simple inspection reveals that they do not last for the same amount of time. But, by computing the average duration of recession one finds—perhaps amazingly—that it is almost 7 years, the same period of famine inflicted on Ancient Egypt!

---

[2] Superstition on seven claims the number of wonders in the ancient world, the Wise Men in classical Greece, the days of the week, the deadly sins and several other cases. But there are more recent baits for the modern appetite: The global crisis sparked panic beyond control when the NY Stock Exchange plunged by 7.777 % in September 29, 2008, as quoted by Mandelbrot, *The (mis) behaviour of Markets* (2008, p. xi). To face the crisis, the US administration injected $7.77 trillion in banks' recapitalisation, according to *Bloomberg* (28 November 2011).

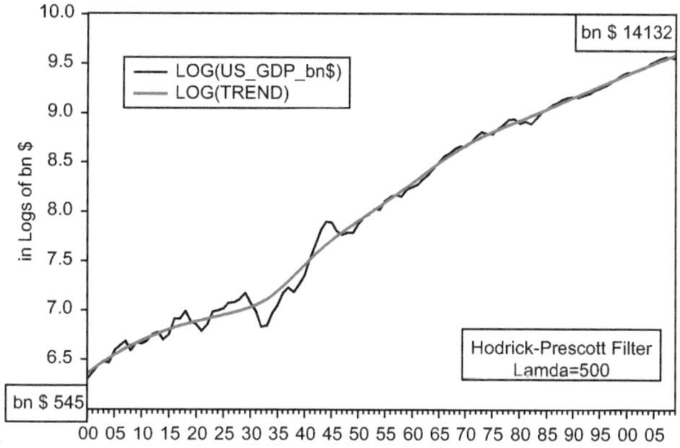

Fig. 1.1 Peaks and troughs in the US economy, 1900–2009. US GDP 1900–2009 and GDP trend, in logs. Billion US Dollar 2000. The trend is obtained thorough a Hodrick-Prescott filter, with smoothing coefficient λ = 500

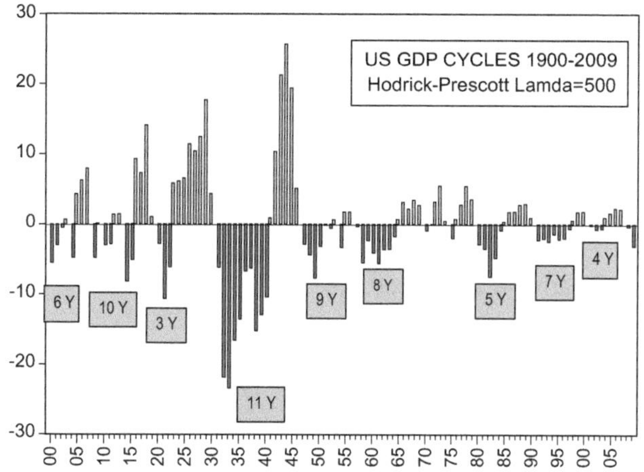

Fig. 1.2 The duration of peaks and troughs in the US 1900–2009. Peaks (above zero) and troughs (below zero) in US 1900–2009, expressed as percentage of GDP trend. *Grey boxes* show the duration of recession in years (ϒ)

So is there really a mysterious force that since thousands of years guides economic malaise to last roughly the same length of time? Some will rush to believe this, perpetuating the superstition on numerology. They may also find other cases of a similar duration and rush to take it as an obvious confirmation. Such assertions might attract some attention in light social conversation, but they commit all types of errors that a "theory" can suffer from. The "7 year" coincidence should not detract us from noticing some fundamental falsifications of an alleged uniformity in economic behavior, such as the following:

(i) There is no resemblance between the origins of the two crises. In Ancient Egypt the cause was the flow shortage of the Nile, due to the lowering of water levels in the lakes of Western Africa. In the twentieth century the causes stemmed from technology, wars, stock exchanges, and also from a persistent effect curiously phrased as "diminishing marginal returns of factors of production".

(ii) The mere replication of duration in Pharaoh's dream and twentieth century US crises does not establish any kind of forecasting. Moreover, if one takes into account how many times the expectation of such coincidences proved to be wrong, even the most fanatic supporter of numerology will be shaken.

(iii) Although the two phenomena are completely different and the attempt to present them as similar is intellectually wicked, it does not mean that analyzing the consequences of recessions in different eras is irrelevant for today. The Rhine may not be at risk as the Nile was in the ancient past, nevertheless Pharaoh's policy in response to the crisis might, even today, tell us something important whether positive or negative.

(iv) The duration of each crisis is not predetermined—let alone at a fixed interval—but it varies according to the type and intensity of its cause and the policies applied in response. For example, the famine in Egypt might have been considerably shorter, had the Pharaoh avoided extensive sales of grain abroad to enrich his Treasury. Analogously, the consequences of the Great Crash of 1929 in the US might have ended earlier, had the policy makers not been so adamant on implementing strict monetary policies. As a matter of fact, some of the Nobel Prize winners present in the meeting—such as Joseph Stiglitz and Eric Muskin—pointed out to Mr. Schauble that excessive fiscal austerity is more likely to prolong the crisis, and, to overcome it, alternative policies should be pursued. Regarding duration, Robert Mundell suggested that structural adjustments could have been completed within 2 years and then growth would resume.[3]

Motivated by the unexpected coincidences regarding the famine in Ancient Egypt, the US recessions in the twentieth century and the first major global crisis of the twentyfirst century, one can tentatively draw comment regarding economic perceptions, their historical origin and how they are used or misused: superficial similarities in economic phenomena do not imply that the latter can be explained by the same theories. Similarities are found in many phenomena in Nature which are governed by the same laws (such as the law of gravity), but economic phenomena are critically determined by social factors which are more likely to be different in each period.[4] Superficial description of an economic event should not be taken as "a theory". Such a claim can be made only if it can explain the origins of the event and is capable of analyzing similar ones observed under different circumstances. Predictions should not only be noticed when they appear successful, but also when they fail to predict the actual facts. Adopting a theory to examine an economic event and guide the response to its consequences should be done with care, because if later is found to be false it may be impossible to repeat and correct it retroactively.

[3] Joseph Stiglitz is the NLE of 2001, Eric Maskin of 2007 and Robert Mundell of 1999.

[4] In an extreme rejection of the forecasting ability of theories, Wittgenstein warned that '. . . future events cannot be discerned from past ones. A belief in deterministic structures is a superstition'. *Tractatus Logico Philosophicus* (1978, p. 35).

Unlike laboratorial experiments which can be repeated at will, social dynamics occur just as unique historical episodes.

But even a detached approach is far from being neutral. Actual events become conflated with policy outcomes thus the assessment of a particular theory is a very challenging task. In practice, it is likely that whenever a theory explaining an economic crisis is examined, it also encompasses the prescription of facing the consequences of that crisis. Having the above caveats in mind, a critical examination of economic theories in historical context is attempted in this book, focusing on how they were generated and influenced by preceding crises.

1.2 Is the History of Economic Thought Any Useful?

One who is concerned with current economic issues may in principle wonder why attention should be paid to the history of economic thinking. Isn't it enough to study contemporary theories and—by employing some sensible criteria—choose the one that appears to be best suited? What benefit is there in learning about economic thinking in other eras and societies, let alone in those of the antiquity such as Greece, Rome, Egypt, Asia or even the Mayas in ancient America?

An answer might be that the history of economic thinking is useful not only in acquiring historical knowledge and understanding of how previous societies have developed or perished, but also helps to reflect upon current economic problems. Several facts that with a sense of excessive self-esteem are considered as unique in our generation, have had already happened in the past in a similar manner. As of today, each epoch attempted to understand them by its own means and intellectual capacity available, thus it is interesting to know how these remote events had been interpreted, and then compare current behaviour and responses with those prevailed in other societies. For example, by which policy did classical Athens avoid famine while Egypt continued to face food shortages? Why ancient Rome had not nurtured production technology, even though advanced technical expertise was available in several domains? Why did the Industrial Revolution take place in England, and not in richer and more powerful France, or even in China where several advances in technology had been achieved by that time?

It is also useful to know how different people have interpreted economic phenomena in different eras, what use they made of their potential benefits and how they did or did not avoid adverse ones. For example, stock markets collapsed around the same time -and roughly for the same reasons- in London and Paris in the eighteenth century but consequences were catastrophic only for France because each government responded differently. Why governments took so long to understand the crisis in 1929 and respond appropriately? Why the centrally-planned economies failed so manifestly to reach a balance between demand and supply and collapsed in 1989? There were economic and social factors that influenced the above events and is crucial to know which theories were more successful in identifying and promptly recognising their effects.

1.3 The Historical Divide

There is no doubt that the foundations of modern economic thinking—at least as the subject matter is perceived in the Western world—were laid in 1776, when Adam Smith published *The Wealth of Nations*; see Smith (1986). The holy triad of classics was completed by Robert Malthus and David Ricardo, and later followed by John Stuart Mill and challenged by Karl Marx. By the end of the nineteenth century three major pillars of economic theory were established, namely the marginalist revolution, the historical school and the institutional school. Then came the domination of Keynesianism in the 1930s which was effectively challenged by Monetarism in the 1970s; intense debates between them produced the so called '*neoclassical synthesis*' that constitutes the main core of mainstream economic theory today.

This synoptic description leaves out several previous developments in economic thinking and also ignores the contribution by non-western civilizations and pre-modern cultures. To avoid some of these shortcomings, the book will start by examining how economic views where influenced by various crises in antiquity and then attempt to describe how the formation of modern theories gradually took place. To simplify presentation, a division of historical time is proposed first.

There are several criteria to periodicise the historical process, sometimes objectively based on major events and changes, other times on more subjective ones prompted by political or national developments. When such criteria broaden to include a variety of seemingly critical factors, separation is complicated and analysis turns cumbersome. To mark a historical divide in economic structures and the concomitant evolution of economic thought, two criteria of economic organisation are considered here: the *scale of economic activity and the freedom of making economic choices*. The first concerns the size of production and distribution activities; the second refers to the degrees of freedom in employing factors of production and selecting market goods. The extent to which these criteria are applied in different eras can be used as a criterion to mark the time milestones of economic history for the following reasons:

(a) *Scale of economic activity*: In the long course of economic history, omnipresent is local organisation based on small units of production and ensuring self-sufficiency. Access to the production unit takes the form of private ownership, social property or even hereditary forms of serfdom. The type of access notwithstanding, the small size deters economies of scale so that total production cannot be greatly advanced. This led to the perpetuation of traditional means of production, since the cost of adopting new technologies was prohibitive for small scale producers. Labour mobility was small or non-existent as cultivation was transferred from one generation to another. Local markets were functioning mainly for the surplus that remained after self-consumption and tax paid to the ruler were subtracted.

On the other side is production and, as a consequence, the distribution of goods on a massive scale, geographically as well as in quantity terms. It is due to this type of organization that major changes were entailed in order to take advantage of the economies of scale, either peacefully or by force. In the course of history, such changes mainly took place by the expansion of global trade, the creation of the banking system and—above all—the Industrial Revolution with all subsequent ramifications to mass production and distribution.

(b) *Degree of mobility and choice*: The change in scale often coincides with, or is facilitated by, the shifting in the forms of ownership and control. It is thus important to examine the possibility of freely selecting the type of productive factors, the degree of their utilisation, and the place on which they are employed; also the degrees of freedom that producers have in distributing their products and that of consumers in choosing the final goods. The extent to which this criterion is found to be applied in practice determines whether the main economic relations in production, employment and exchange are set by command in predetermined levels or are influenced by market forces.

In economic history, the transition from one mode of production to another was determined—to a great extent—by the radical changes in scale as well as in factor mobility. For example, the feudal mode was set up by enforcing the villagers to stay *in-situ* and cultivate holdings that were expropriated by the new rulers after the fall of the Roman Empire. When restrictions on trade were relaxed, commerce expanded vigorously and, combined with the new frontier of production possibilities, shook feudalism to extinction. When labour ceased to be ruled by landlords a more tolerant regime of job searching prevailed thus enabling large-scale employment during the Industrial Revolution.

The concepts of production scale and factor freedom are also found to be critical in several episodes of economic crises and collapse. Studying the reasons behind the sudden disappearance of mighty civilizations, Joseph Tainter argues that a society is collapsing "when it displays a rapid, significant loss of an established level of socio-political complexity".[5] As complexity rises with the scale of production, the ramifications of a shock that impinges upon an economy may be more catastrophic if a critical threshold is surpassed beyond which its effects are no more manageable. At a lower level of a shock, however, the scale of activity may offer an advantage on alleviating the consequences in a less painful way.

Freedom of economic factors is crucial too in the extent and impact of a crisis. For example, the fewer the rigidities in a labour market, the faster an unemployment shock is expected to wither. Although central coordination is usually indispensable in fighting a crisis, this must not go as far as to deter individual and decentralized choices on how best they react to adverse conditions. In a highly influential diatribe on the links between oppression and crisis suffering, Amartya Sen explained why famines in China and the Soviet Union during the twentieth century cost infinitely more lives if compared to similar phenomena that took place in more open, though sometimes poorer, societies.[6]

[5] From Tainter, *The Collapse of Complex Societies* (1988, p. 4).

[6] Sen, *Development as Freedom* (1999, Chap. 7, Famines and other crises, pp. 160–188).

Patterns and consequences, of course, are not always one-sided and uniform. If, for example, consumption is freed from a state-rationing system and relies on individual choice, it is most likely that new markets will emerge leading to more elaborate economic structures. But occasionally it may also lead to extensive shortages for those deprived of the means to buy these goods, as happened for example after the markets of essential goods were liberalised in post-communist Russia.

In the course of history, the shifting of control of production in the event of crisis may also go in the opposite direction of less factor freedom and toward more public ownership and state management of distribution. Economies of scale change too, but again results are not one-sided. For example, it turned out to be the case that some activities became economically feasible only after nationalization of the means of production, as happened in the steel industry in mid-war Soviet Union or in several public utilities in postwar Western Europe.

With the above caveats in mind and in order to avoid a detailed historical phasing, the evolution of economics can be classified into two periods: the 'pre-modern' era considered to span from antiquity to the end of fifteenth century and the 'modern' era from 1500 onwards. The late fifteenth century witnessed major historical events that profoundly changed economic structures worldwide: the fall of Constantinople in 1453, the consequent Renaissance in Italy, the inwardness of China after abandoning distant seafaring in 1477, or the waves of Church reforms proved to be landmarks in the march of history.[7] Above all it was the discovery of America that made 1492, "the best year from which to date the beginnings of the world we are in now and the era we call modernity".[8]

Of course, other alternatives could have been suggested. For example, Graeber (2011) confirms that the late fifteenth century "was marked by so many things that were genuinely new—the rise of modern science, capitalism, humanism, the nation state'; nevertheless he notes that in the systems of exchange a backlash took place "from virtual currencies and the credit economies and back to gold and silver".[9] Clark (2007) suggests the milestone of year 1800 after which incomes differed sharply between various countries depending on whether they endorsed the Industrial revolution or failed to do so.[10] But from the point of view of economic theory, the social and political factors that fermented the conditions for the industrialisation of production can be traced from long ago. Scale and freedom vastly different before and after the fifteenth century as explained below:

[7] Hardly can one consider such events in isolation. The vacuum created by the Chinese withdrawal and the threat imposed by the rising Ottoman Empire on the naval routes made the discovery of alternative westward access to the East all the more pressing. For trade to thrive, the prohibitions of the Church should be eased.

[8] According to Fernandez-Armesto, *1492: The Year the World Began* (2009, p. 12).

[9] Graeber, *Debt: The first 5,000 years* (2011, p. 308).

[10] Clark, *A Farewell to Alms: A brief economic history of the world* (2007, pp. 1–16).

1.3.1 Pre-modern Period: Small Scale of Production—Low Factor Mobility

From antiquity until roughly the end of the fifteenth century, economic life was characterised by a high degree of control and many constraints on production factors, resulting in very limited mobility, if any at all. The most characteristic facts of this long historical period are the various forms of slavery and serfdom as the main ways to meet labour needs in production. But even if not taking the form of slavery, employment was nevertheless characterised by a long fixity, effected by adhering either to family holdings (as in farming) or to vocational tradition (as in the guilds). Commerce was also severely curtailed for a long period in medieval Europe, not enabling the free exchange of goods and the distribution of resources.

The pre-modern period was dominated by small scale production and characterised by a thin accumulation of capital, though there have been some noticeable exceptions. For example, small cultivations were eclipsed in ancient Egypt when encompassed into a system of state management, so that the economies of scale in irrigation offered by the Nile were exploited. Small farmlands were also merged to form the large *latifundia* in the Roman age so as to take advantage of organized slavery and the access to the transportation infrastructure of the Roman highways.

The forms of economic thinking that appeared in the pre-modern era differ from those in the modern one regarding structure, extent and purpose, precisely because they were motivated by highly different economic conditions and scale. Most of the issues concerned small production, local commerce and in-kind taxation, and were usually expressed as just an advice to sovereigns in order to secure the stability of supply—and their own regime. For precisely this reason, economic thinking typically takes a 'didactic' form on how things *have to happen*, and rarely as systematic observations of what *does really happen*. Even the greatest of pre-modern thinkers—such as Aristotle in classical Athens, Fan Li in ancient China, or Haldoun in Islam were driven by wider philosophical considerations aiming to inspire a prudent individual conduct or a certain state function, and pay less attention on how to explain economic phenomena or organise them in a knowledge structure that could be tested in practice.

1.3.2 Modern Period: Large Scale of Production—Increased Factor Mobility

The new period dawns around 1500 when the scale of activity and the degree of freely choosing and moving production factors increases significantly. The vast rise and expansion of commerce, the free movement of villagers into towns, the accumulation of wealth and the appearance of new techniques in production did not remain isolated phenomena. They were combined with sweeping demands for freedom and human rights, against the absolutism of the previous period.

Soon this wave leads to new economic structures, and new theories appear attempting to explain the consequences and risks they bear. They fervently seek answers to where society will be driven, and whether the outcome can be more beneficial for society than before. The milestone of the new era is the Industrial Revolution that takes place two centuries later and coincides with the birth of modern economic thought.

In an era when society is gradually freed from repression and various constraints, it is difficult to reach consensus on how the '*common good*' should be defined, and this drives economic theories to take sides by defending or disavowing social classes and their interests. Hence, a new characteristic of modern economic theories is their mutual rivalry to a degree previously unknown when economic thinking was mainly encouraged and monitored by the ruler. This is a mixed blessing for the newborn field: On one hand, it exerts a ceaseless motivation to improve upon other theories, but on the other it also means that there never will be a universally accepted single theory for interpreting or influencing the economic process.

1.4 Plan of the Book

Chapter 2 lays out the various forms of economic theories and crises. There are various types of economic thinking, ranging from practical rules to complicated theories, according to the need they are purported to serve. Crises are classified as supply-side or demand-side shocks, and the latter are further detailed according to the type of asset or income that causes the disturbance.

Chapter 3 starts examining some economic functions and crises in the ancient world. From the 7-year famine in ancient Egypt to the decline of classical Athens, crises spark the search of explanations, policies and organizations to face them, gradually leading to the formation of economic theories.

Chapter 4 examines some economic crises in the Roman era, from an early real-estate bubble to the monetary debasement and inflation that undermined the last phase of the Empire. To avoid such crises and revive the economy, the Eastern Roman Empire adopts more efficient rules and practices, in contrast to the economic decline that befalls in medieval Europe.

Subsequently, Chap. 5 examines the economic practices in medieval Europe and how they were shaken by famines and tax revolts, before new theories start to provide guidance for the emerging modern era.

Chapter 6 examines the origin of economic theories before the Industrial Revolution and their rivalry as they try to defend different sectors in society. Financial crises erupt in the most developed economies, and new questions emerge on how they are best dealt with.

Chapter 7 elaborates on why England was best prepared for the Industrial Revolution versus other economies in Europe or elsewhere. The foundation of classical economics follows with the triad of Adam Smith, Malthus and Ricardo.

Chapter 8 examines how the crises after the Industrial Revolution led to social discontent and the formation of Marxism as a radical new economic theory. As other major economies enter the industrial era, economics are further developed and new questions arise.

By the late nineteenth century, attention of economic theories is shifting from accumulation to distribution, and this is examined in Chap. 9. Utilitarianism appears as the new organizing framework of society, while economic theories are faced with repetitive crises and search for how equilibrium can be reached.

Chapter 10 describes how the Great Crisis in 1929 incapacitated most of the prevailing economic theories and paved the way for the new theory of Keynesianism and introduced large-scale quantification in economics.

The twentieth century witnessed a wholly different economic system based on Marxist theory and Central Planning to be widely established. Chapter 11 examines a series of crises and how mixed theories tried to correct its failures, until it was ultimately collapsed.

Chapter 12 describes the turn of Keynesian economics to be challenged after failing to apprehend the stagflation crisis in the Western economies. Monetarism appeared to dominate for some time, before a more constructive synthesis between rival theories finally emerged.

As development became the main postwar inspiration worldwide, new theories appeared as alternatives among Western capitalism and Eastern central planning. Chapter 13 examines how most experiments ended in crisis and new approaches were devised to face them.

Chapter 14 describes how the global crisis in 2008 has shaken the foundations of dominant economic thinking and puts forward the challenges for new theories and answers.

Finally, Chap. 15 discusses some methodological issues in economics and how studying economic history and the history of ideas can help to improve our understanding on crises and theories.

References

Clark G (2007) A farewell to alms: a brief economic history of the world. Princeton University Press, Princeton

Fernandez-Armesto F (2009) 1492: the year the world began. HarperOne, New York

Graeber D (2011) Debt: the first 5,000 years. Melville House, New York

Mandelbrot B (2008) The (mis)behaviour of markets: a fractal view of risk, ruin and reward. Profile Books, London

Sen A (1999) Development as freedom. Knopf, New York

Smith A (1986) The wealth of nations. Penguin, London

Tainter J (1988) The collapse of complex societies. Cambridge University Press, Cambridge

Wittgenstein L (1978) Tractatus logico-philosophicus. Cambridge University Press, Cambridge

Forms of Economic Theories and Crises 2

Abstract

Interactions between economic crises and economic theories have been strong and important, as the authority in charge of the economic entity (be it the ruler, the state or an international organization) tries to respond to it by seeking policy guidance and explanation. Economic theories fall or flourish, adapt or perish, according to whether or not they are capable of facing this challenge. The outbreak of a crisis may be the reason for a previous tenet to vanish or the opportunity for a new one to be born and prevail. An economic theory that is well articulated may lead a society to prosperity and stability by avoiding major crises or—if it is frail and erroneous—to bring about instability and collapse.

2.1 The Relationship Between Crises and Theories

Economic history and the history of economic thinking are in general intertwined, since economic events are an input necessary either for testing or inspiring an economic theory. The reverse is also true though not always admitted. When, for example, economic agents make decisions on how long to work or retire, and households on how much to consume or invest, they are always under the influence of certain perceptions and ideas, without necessarily knowing that they follow a particular economic theory. The interplay between theories and crises is seldom clear-cut for a variety of reasons:

2.1.1 Organic Theories

Economic theories influence actual developments more directly when decision-makers consciously take them into account in order to determine their actions or expectations. In this case, an economic theory is said to have an *organic relationship* with the economic process. For example, Keynesianism used to be the

N. Christodoulakis, *How Crises Shaped Economic Ideas and Policies*,
DOI 10.1007/978-3-319-16871-5_2

operational economic theory for post-war Western Governments employing expansionary fiscal policies to face recessions. Monetarism was in the 1980s the operational economic theory for the then Government in UK and the Administration in the US, who strictly controlled the money supply in order to fight inflation. Marxism was the organizational theory in the centrally planned economies, in which decisions on setting prices were justified by appealing to the labour theories of value as developed by its founders.

2.1.2 Amendments and Insistence

Sometimes the founders of a theory are personally involved in the conduct of economic policy, and in these cases it is easier to identify a particular theory with the policy being implemented. However, in the occasion of major deviations between theoretical assertions and outcomes, two different mending actions might be undertaken [1]:

First, the initial theory may be adapted by scrapping or moderating some early convictions so as to take into account actual developments. When, for example, Lenin in the early years of the Soviet Union recognised the immense difficulties in running a real economy as opposed to ideological sermons, he dropped the early obsessions of completely eliminating the price mechanism. Another example was the admission by mainstream economists in the 1980s that the Phillips Curve may not always have a downward slope and, therefore, the effectiveness of a monetary expansion to tackle unemployment can become totally ineffective.

Second, in authoritarian societies the interpretation of reality may be adapted in such a way as to conceal the failure of a theory. Economics is not the only science suffering from this problem, and Karl Popper has memorably criticized those who believe they hold the keys to absolute knowledge and harmony: from the naturalist philosophers of the sixteenth century who believed that everything is explained by the laws of mechanics, to the proponents of idealism who tried to discover the "elegant harmony" of events and held doctrines based entirely on untested ideas.[2]

This attitude is often encapsulated in the motto *if the world does not comply with the theory, so much worse for the world.*[3] This reflects a combination of voluntarism toward reality and cynicism toward society. Outstanding examples have been the

[1] Economic history is full of cases where a Government claims to adhere to a specific economic theory, but the policy that is actually applied seriously deviates from what should have been expected. For example, the US Administration of G. W. Bush was preaching the curtailment of public expenditure, but in 2008 public debt and the twin deficits were at record levels. The Communist Party in China endorses Marxism-Leninism as its guiding theory, something that is hard to believe by seeing the number of billionaires to rise by the year and inequality to widen even further.

[2] Popper, *All Life is Problem Solving* (1999, Chap. 15, pp. 341–349).

[3] The aphorism is attributed to Hegel, though no written statement by him is found. Such views were endorsed by Jesuits who preached that knowledge should remain subordinate to Church's

official silencing on the famines raging in China in 1958 or in the Soviet Union a decade earlier, in the calculation that the theory of central planning would remain immune to public anger.

But concealing the true consequences of a policy hardly was the monopoly of Marxist regimes. A stubborn claim by monetarists was that the outbreak of unemployment in the 1980s would be transient and dissipate as soon as inflation is put under control. Ten years after the experiment and with unemployment in double-digit levels, the proponents of the theory offered that the reason behind its persistence was not the inadequacy of their approach but the imperfections in the labour market.[4]

2.1.3 Scale and Relevance

Managing a crisis often requires radical shifts in the form of economic activity. Sometimes by imposing more centralized structures, to achieve economies of scale such as in the so called 'Asiatic mode of production' in ancient empires, or the intensive state intervention during the 1930s to compensate for ineffective aggregate demand. At other times, it happens by crushing the centralized structures and liberalizing markets as witnessed in Eastern Europe in the 1990s.

When a crisis exceeds the borders of a particular national formation, its management requires the setting up of new institutions with a supranational mandate to control the consequences and remedy the fundamental causes. A major institutional design in the aftermath of a crisis took place in the post-war period by designing new exchange rate regimes (through the Bretton Woods system), boosting international trade (through World Trade Organization) and encouraging large economic unions (such as the European Union in the western or the Comecon in the eastern part of the continent).

In the historical process there have been periods in which economic theories remained unaware of the changes in the scale of economic activity and their implications in the event of a crisis. For example, the Physiocratic theories were in line with the scale of self-sufficiency production professed in seventeenth century France or Germany, but were not able to understand and manage the wave of industrialization and the rapid expansion of world trade. Until the Crash in 1929, "no general theory of how and why [. . .] crises occurred had emerged".[5] The neo-Marxist theories of underdevelopment and unequal exchange failed to secure the conditions for prosperity and inward development in Latin America in the 1960s, at a time when more international integration was needed.

belief; see Ignatius Loyola, *The Spiritual Exercises* (1914, Rules to have the True Sentiment in the Church, Rule 13).

[4] Monetarism started to be implemented in UK in 1979 and in more intensive form in 1983. Ten years later, in 1993, unemployment in UK was above 10 %. For a discussion see the debates in Booth, *Were 364 economists all wrong?* (2006).

[5] In the words of Wolff and Resnick, *Contending Economic Theories* (2012, p. 317).

A similar decoupling between dominant theories and their capacity to understand the economic environment happened recently in relation to the global crisis in 2008. Economic theories were in a process of perpetual refinement by describing in increasing detail the behaviour of "representative agents", such as the household, the firm or the Government, and examining under which assumptions and constraints their decisions can lead to the best possible outcome. In the meantime, the relentless liberalization of economic borders and the increasing mobility of factors caused economic activity to move from a national framework to a far more globalized one. Economic theories were unprepared to explain the consequences of fast globalization on the stability of the economic system, a concept hitherto analyzed only in a national context. When the tide went too high, the result was governments and national central banks to lose any sort of dirigisme and plunge into global trouble.

2.2 Forms of Economic Thinking

Though differing in structure, extent and impact, interpretations of economic developments have appeared, one way or another, in all epochs. So the immediate question is when exactly can we talk about an *economic theory*? Lionel Robbins (1898–1984) has famously summarized the analysis provided by economics as consisting of

> deductions from a series of postulates, the chief of which are almost universal facts of experience present whenever human activity has an economic aspect, the rest being assumptions of a more general nature based upon the general features of particular situations . . . which the theory is to be used to explain.[6]

It is impossible to find that all economic beliefs and reasoning satisfy the above definition. Designing a way to measure the output of agrarian slaves in antiquity does not constitute a 'theory', even if the method is based on extensive trials and identification. Advising the rulers to accumulate bullions of gold in order to avoid the debasement of coins is a worth noting economic practice that could have been taken as a forerunner to hedging but, without explaining why such things happen, it hardly can be treated as a 'theory' either. For these reasons, it is better to view historical economic thinking as a spectrum of perceptions, propositions and reasoning, sometimes thorough and far-reaching, at other times superficial and narrow-minded. The various approaches can be categorised into four types as follows:

(a) *Prescriptions*: include measures, laws, commands and rules devised to determine the economic conduct of individuals and societies, but they lack a wider philosophical or scientific foundation.

[6] Robbins, *The Nature of economic generalizations* (1979, p. 46).

(b) *Descriptions*: include comments, thoughts and observations that are made by historians, travellers, intellectuals or even rulers themselves on various economic phenomena. More often than not, they are a collection of unconnected essays rather than a coherent and integral set of thoughts.
(c) *Admonitions*: It is a comprehensive combination of (a) and (b) above, aiming to influence decisions and guide behaviour. To support their arguments, the authors try to underpin them as a consequence of other established systems (mainly philosophical principles, religious doctrines, or mathematical laws). For example, the denouncement of interest-bearing loans by medieval Church or by all-time Islam was justified by recalling the preaching of their founders, so as to make the argument more persuasive for the populace and more threatening in case of trespassing it. The most famous example in admonitory political philosophy is *The Prince*, written by Nicolo Machiavelli in the sixteenth century. The *Arthasastra* written in ancient India, the texts by Fan Li in ancient China, or certain parts of the *Summa Theologica* by Thomas Aquinas in the thirteenth century are close analogies of economic counselling.
(d) *Theories*: These are the most integrated approaches on economic matters, by structuring their principles in a systematic way and allowing their conclusions to be tested empirically so that they can be used to explain and/or foretell economic developments. Later in this book, some criteria on which an economic theory can be more thoroughly assessed will be discussed.

2.3 A Typology of Crises

To examine the interaction between economic theories and crises, a typology is required for the latter as well. In the first place, economic reasoning stemmed from a situation of scarcity where the goods or the resources that were pursued were in excess of those actually available. This was famously summarized by Robbins that "economics is the science which studies human behavior as a relationship between ends and scarce means which have alternative uses".[7] Taking this concept a step further, we can define a situation as being a crisis if the degree of scarcity is abruptly multiplied, leading to major imbalances. This approach is helpful in classifying economic crises according to their cause, as follows:

2.3.1 Supply-Side Crises

Massive and abrupt disruptions in the production or supply of goods aiming to cover extensive and essential needs of a society led to several economic crises. The oldest and most frequent crises are the famines, caused by acute shortages in food supply, usually grain. Famines are caused by the destruction of cultivated lands

[7] Robbins, *An Essay on the nature and significance of Economic Science* (1935, p. 15).

(due either to droughts or floods), by epidemics killing farmers or livestock, and also from extensive corrosion of land due to chemical warfare, nuclear radiation, or industrial pollution. Frequently a famine is followed by administrative breakdowns in distribution and production management: as a result the crisis is multiplied leading to human catastrophes. Compared to the pre-modern era, technological progress in cultivation and breeding radically reduced famines in modern times.[8]

Other supply-side crises are due to shocks on input prices, such as the price of oil in the 1970s, or the collapse of production factors such the obliteration of factories in World War II, or the extermination of skilled human power (as purging the Huguenots from France in the seventeenth century or Jews in Nazi-occupied Europe). More recent examples have been the genocides of populations in oppressive regimes or the industrial catastrophes in Chernobyl in 1985 and Fukushima in 2010.

2.3.2 Demand Crises

They are caused by a sudden and massive fall in the means through which consumption is financed, most commonly financial or real assets. These may include incomes, real balances, bank deposits and any type of wealth assets with a possible transaction use. Demand crises are further classified by the specific way the asset value is eroded, and the most common forms they take are the following:

2.3.2.1 Money Debasement

When a sovereign entity is not able to cover its financial obligations through taxation or confiscation, it resorts to money creation that typically leads to price increases and hyperinflation, corroding the purchasing power of monetary assets and nominal incomes. In the modern era, the debasement of money usually takes the form of issuing notes on a massive scale, as happened in Germany in 1922, several Latin America countries during the 1980s, Russia in 1998, or nowadays Zimbabwe.

Obviously, overprinting was not an option available in pre-modern times, as fiat money did not exist then. But this hardly deterred the imagination of rulers. To increase money circulation, authorities simply had to counterfeit coins by amalgamating them with cheaper metals while keeping nominal value the same as the original. Another method of money debasement is through exchange rate depreciation, where inflation breaks out due to the fast deterioration of the terms of trade. Again, this method is used mainly in the modern era when fiat money circulation and specific exchange rates are established. In the pre-modern period

[8] Though they did not totally disappear. Famines continued to occur in Asia, Africa, or even in relatively more developed societies, such as Ireland (1845), Greece (1942), the Soviet Union, and elsewhere. For a brief description see O'Grada, *Famine: A short history* (2009).

the rate of coin exchanges could also be depreciated when the currency of one country was found to be counterfeited.

2.3.2.2 Banking Collapses

Monetary assets may collapse not only through hyperinflation, but also if the financial institutions where they are kept go bust. Looking at the balance sheet of a bank, one can distinguish two types of such crises according to weather they involve bank's liabilities (retrievable deposits) or bank's assets (collateralized loans): A bank liability crisis occurs when there is a run to the banks by panic-stricken depositors, as happened several times in Europe in the nineteenth century and in the US in the early 1930s. A bank asset crisis occurs when the repayment of the bank's loans to third parties is seriously undermined, as happened with the toxic mortgages in the US in 2008.

In both cases, the crisis is triggered by an imbalance between lending and borrowing. Naturally, such crises occur only after banks were established, and the first banking crisis is recorded around mid-fourteenth century when England failed to repay loans from the main banks of Florence, and two of them collapsed. From this angle, major clashes between lenders and borrowers in antiquity could also be viewed as 'banking crises'. It was for that reason that a great deal of economic thinking of the time was devoted to the treatment of interest and the conditions of obtaining and repaying a loan. In the ancient world and also later in the Roman and Byzantine periods, interest was permissible within certain limits, while the inability to repay a loan was frequently punished with debt bondage. In ancient India credit was legal, and interest rates were flexible up to a level of 15 % annually; in Western Europe interest was forbidden to Christians until the thirteenth century, while in Muslim countries it remains so to date.

2.3.2.3 Collapse of Financial Assets

A crucial characteristic of modern-era economies is the creation and vast expansion of capital markets, bond markets and derivatives markers. Through them economic agents allocate their assets along a time span and across alternative options, in such a way as to maximize their returns under a number of assumptions. Since outcomes are uncertain these allocations rely on 'expectations' of future returns, and thus they are hit whenever predictions fail. The more abrupt the failure of expectations and the tighter the connection between future returns and current demand, the more extensive the ensuing crisis will be. Two are the most common forms:

Stock market crises, in which future dividends are dwindled and an immediate undervaluation of the invested capital takes place. A typical crash occurs when expectations of future returns are multiplied as a 'bubble' that eventually collapses as no more supported by real economy fundamentals. It is interesting to note that bubbles may occur not only as a result of false information on future returns, but also as a rational bet as long as there are investors still willing to believe that misinformation may last in the next period and a profit can be made from it.

Bond market crises, when the borrower is unable to repay all or part of debt obligations, thus causing an equal fall in the value of bonds held by the creditors. As bonds are to a large extent issued by sovereigns, crises occur when a state faces default. The worst case of default is when debt obligations are owed to foreign creditors, in which case the country is pressed by the international credit system or even by other Governments acting on behalf of their investors.

In contrast, when a Government debt is denominated in domestic currency, it seldom leads to default and collapse, since obligations can be met through money creation to nominally repay bondholders. In this case, a demand-deficiency crisis may occur through hyperinflation caused by the extensive increase in money supply. Otherwise, domestic debt may lead to a major economic crisis in case the sovereign is deprived from issuing new money, as happens today in the Eurozone countries.

2.3.2.4 Collapse in the Value of Real Assets

The most recurring example is the real-estate market. The rise in housing prices caused by higher demand augments the value of the asset and, as a result, raises current consumption. But the process of rising prices may come to a halt if the financing of acquisitions is abruptly cut or the attraction of existing real-estate falls. The collapse of housing prices leads to deficient aggregate demand and recession, as happened in Florida, US, in 1920–1926, Japan in the 1990s, and the US and Europe in 2008. In the pre-modern era, a real-estate crisis took place in ancient Rome in the years right after the civil war.[9]

Similar crises stem from other types of real assets being depreciated, such as product inventories, art collections, stocks of precious metals, etc., though the recessionary effects are usually milder compared with real-estate crises. The most famous example is the bubble on tulip prices in the Netherlands during the seventeenth century, and since then '*tulipmania*' is a term frequently used to characterize market collapses. Less known is the depreciation of spice inventories due to the falling prices of pepper after the navigation route around Africa was established in 1503.[10] Another type of asset collapse took place in modern Europe when the value of art collections of Jewish painters (such as Modigliani, Sagal, Pissarro, Liebermann, etc.) fell dramatically after Hitler assumed power in Germany in 1932 and fiercely attacked "degenerate art".

[9] As described by Suetonius, *The twelve Caesars*, (Book A, XLII) and discussed in Chap. 4.

[10] The navigation around the Cape of Good Hope led to a substantial reduction of risks and transportation costs from Asia to Europe, and resulted to much lower prices of the merchandise. Between 1500 and 1525 AD, the price of pepper had fallen by 70 % in real terms, as shown in Findlay and O'Rourke, *Power and Plenty: Trade, War, and the World Economy in the Second Millennium* (2007).

2.3.2.5 Ramifications

Frequently a crisis is not one-dimensional but transgresses other sectors of economic and social life, and thus its consequences are viciously multiplied. For example, by depreciating a household's wealth and savings, a real-estate or a stock-market crisis soon affects the banking sector as well. A natural catastrophe initially causes a crisis of supply, but it also precipitates a demand crisis as the incomes of producers are lost.

Nor is the propagation of a crisis unidirectional for all sectors. For example, the exchange rate crisis and default in Argentina in 2002 led to a massive reduction of real wages and under-consumption, but stock market prices skyrocketed.[11] In a symmetric way, the stock market plunge in 2000 following the collapse of dot.com worldwide, made bond markets a lot more attractive and sovereign paper valuations thrived. However, during the 2008 crisis both markets were hit badly.

2.4 What Constitutes a Crisis?

2.4.1 Crisis Thresholds

An adverse situation is called a crisis either because the economy has been badly shaken (input crisis) or the adverse consequences of some effect become extensive (outcome crisis). The former case requires the definition of a quantitative threshold (for debt, inflation, external deficit, and so on), while the latter an impact assessment in critical areas, as for example in unemployment, production, reduction or redistribution of income, etc. In a recent book on the classification and analysis of financial crises, Reinhart and Rogoff (2009) suggest that a threshold for inflation is 20 % annually, for an exchange rate crisis a devaluation exceeding 15 %, while for debasement crises a degree of forfeiting more than 5 %.

A prolonged suppression of returns on wealth, due to market curtailment or institutional repression is also considered as a 'crisis' by the same authors. When for example there is a financial repression, households do not have the option of allocating their savings except in low-return deposits, as was the case in the centrally planned economies of Eastern Europe. This amounts to imposing a partial default, though effected quietly and over a long period of time.

2.4.2 Repression

Following the same logic, one can classify long periods of economic standstill or the imposition of low returns on economic agents as a *factor repression*. For example, slavery constituted a long labour crisis that ended with the emancipation

[11] During 2001–2008 the peso to dollar exchange rate collapsed from 1:1 to 4:1, but the over the same time the Stock Exchange index increased thirteenfold. Source: *tradingeconomics.com*.

of slaves, feudalism a long crisis in capital accumulation that ended with the advent of capitalism. Karl Marx considered the very system of capitalism as a long crisis of exploitation that would inevitably end by victorious socialism. On the other hand, Friedrich Hayek thought of socialism as a 'repression of potential' that would end by the liberalization of markets.

2.4.3 Reaction Patterns

Looking through a crisis framework, some patterns related to economic theories are frequently repeated:

Insistence when a crisis erupts, the most common reaction of dominant theories is not to adjust to the new realities but to embark on an enormous task to show that developments and remedies fall within the provisions of already applied policies. Frequently adjustment failures result in a 'positive feedback' reaction that further intensifies the crisis rather than defusing it.

Habit Persistence Sticking with the old way of thinking may be due to the complexity of prevailing theories that increases the cost of adjustment. The more complicated and widespread a theory is, the more difficult is to revise or abandon it in the event of a crisis. The reason is that the *esprit de corps* of current followers deters alternative arguments, fearing that they may be stripped of privileges and access to power.

These remarks help to raise some questions on how theories should be assessed in the face of an economic crisis:

(i) One criterion is the extent to which they can predict and thwart a shock on, or an adverse development in the economy. For example, the interventionist theories during 1950–1970 managed to deal with fluctuations of demand, but failed when the oil price shocks impinged upon them in the 1970s.
(ii) Whether they can foresee the consequences of a crisis and suggest how their impact is minimized afterwards, as happened with Keynesianism in the recession of the 1930s or with monetarism in the stagflation of the 1980s.
(iii) Whether they contribute to the ending of the repression of factors or help to keep the lid on them. The aphorisms by medieval Church against commerce repressed both production and consumption, while the theories of Adam Smith argued for setting capital and labour free sped up the Industrial Revolution.

In the following chapters, the answers to these questions will be traced back to antiquity.

References

Booth P (2006) Were 364 economists all wrong? The Institute of Economic Affairs, London

Findlay R, O'Rourke K (2007) Power and plenty: trade, war, and the world economy in the second millennium. Princeton University Press, Princeton

O'Grada C (2009) Famine: a short history. Princeton University Press, Princeton

Popper K (1999) All life is problem solving. Routledge, New York

Reinhart C, Rogoff K (2009) This time is different: eight centuries of financial folly. Princeton University Press, Princeton

Robbins L (1935) An essay on the nature and significance of economic science. Macmillan, London

Robbins L (1979) The nature of economic generalizations. In: Hahn F, Hollis M (eds) Philosophy and economic theory. Oxford University Press, Oxford

St. Ignatius of Loyola (1914) The spiritual exercises. P. J. Kennedy & Sons Printers to The Holy Apostolic See, New York. Available at ixtmedia.com, The Digital Catholic Bookstore

Wolff R, Resnick S (2012) Contending economic theories: neoclassical, Keynesian and Marxian. MIT Press, Cambridge, MA

How Old Are Economics? 3

Abstract
Similarities between ancient and current malaises are not necessarily found in the form that a crisis erupts, but certainly in the way that a mechanism of reaction, recovery and future prevention is carried out. The 7-year famine in ancient Egypt is a prime example of crisis-management that deeply influenced economic developments in the then known world. Later, a crisis of over-indebtedness in classical Athens led to a wide range of political and economic reforms that proved to be pivotal to the rise of democracy and power. In turn, the decline of Athens prompted philosophers to think how the City should be restructured so as to regain prosperity, and this gave birth to several advances in economics. In other empires, a centralized system was adopted for collecting and distributing production, in many ways foretelling the central planning applications of the twentieth century.

3.1 Economic Functions in the Ancient World

The economy, always and everywhere, performs two fundamental functions: setting up production and carrying out distribution. Whichever economic system we study—be it of biblical Egypt, ancient Greece, eighteenth century England or modern day China—we find out, perhaps surprisingly, that the models around which these two functions are organized are limited to a very small number of choices: communal tradition, central control, and markets, or some combination of them.[1]

Social formations historically emerged as small communities, in which the "*oikos*" (house) was the dominant organizational unit of production and distribution. Economic behavior in those early social formations was not an independent

[1] This remark was made by Heilbroner and Milberg, *The making of economic society* (2008, p. 37).

N. Christodoulakis, *How Crises Shaped Economic Ideas and Policies*,
DOI 10.1007/978-3-319-16871-5_3

activity, but rather a component "embedded" in general standards and needs.[2] Economic rules needed to be easily understood and adhered to, and the most effective way to embed them was through religious doctrines. The mode of organization was mostly based on tradition, with each generation continuing in the same way as previous generations, although some technological changes from time to time transformed the structure and scale of agriculture. When affected by an exogenous shock or faces population pressure, this "customary equilibrium", as called by John Hicks, is transformed to a new economic type.[3]

When communities evolve into cities or become integrated into empires, the need to organize a more complex system of managing and administrating the various phases of economic activity becomes pressing. Economic thinking then faces the challenge of analyzing the complications of exchange which accompany the expansion of trade. One of the most recurring and critical issues was how to organize taxation on a larger scale, and how to use state revenues in order to ensure the survival of the population, the functioning of the system and, above all, the power of the sovereign.

When an economic crisis erupts, managing the way out often leads to groundbreaking changes in the nature of economic activity. In dealing with production and distribution deficiencies which precipitated the crisis, more concentrated structures in the economy are frequently emerging in an attempt to create economies of scale and raise efficiency. In other cases with dysfunctional structures, decentralization and liberalization win the day.

3.2 The First Big Crisis: Famine in Ancient Egypt

Crucial issues in economic organization were dealt with at around the same time and in a similar manner in various areas in Mesopotamia and Egypt, and these similarities are likely to reflect a deeper resemblance in economic structures. For example, the Code of Hammurabi in ancient Babylon—the first recorded legislative work in the history of mankind—included provisions on the inheritance of private property, the imposition and collection of fines and the amount of taxes due by each household according to its wealth and land property. There was also concern regarding the effects of lending thus ceilings were imposed on the permissible interest rates.[4]

Similar rules were applied in Egypt, a country that developed one of the most advanced ancient systems of economic organization and state management. Activity was organized around the river Nile, which is still considered to be Egypt's

[2] Similar findings are encountered in Economic Anthropology, a relatively recent research field examining traditional communities of the present, as described in Chap. 15.

[3] Hicks, *A Theory of Economic History* (1969, p. 14).

[4] For an account of interest since ancient times see Sidney and Sylla, *A History of Interest Rates* (2005).

lifeline. The Nile was crucial for the fertility of the land and served as the trade and transportation route throughout the country.[5]

A major concern of economic organization was the sufficiency of grain production and the efficacy of its distribution so as to ensure subsistence. Egypt was the first country to make plans and build public infrastructure, mainly for grain storage and irrigation. Furthermore, since independent producers and merchants were also allowed to operate in ancient Egypt, lending and borrowing activities were regulated, forming a basic system of banking.[6]

The first recorded economic cycle is considered to be the 7 years of the lean-fleshed cow, caused by a drop in the level of the Nile. Were we to describe this phenomenon in modern terms, we would call it an external supply-side shock. According to the well-known biblical narrative, Pharaoh invited Joseph to interpret a strange dream and Joseph foresaw the 7 years of famine that would follow.[7] Subsequently Pharaoh appointed him as head of the Treasury and responsible for the distribution of grain, despite the fact that he was a Jewish slave. Joseph organized a system of compulsory grain storage, imposing a tax to be paid in kind amounting to 20 % of production and sending tax auditors to monitor collection.[8]

New state silos were built to store the grain, and during the high time of famine he enforced a system of food rationing. Since neighboring kingdoms were not as prudent, they were soon begging to buy grain from Egypt; exports increased state revenue and influence abroad, but reduced the quantities available for the Egyptian population. As a consequence, grain stocks began depleting before the end of the famine and were sold at very high prices. Those who could not afford it were forced to mortgage their lands resulting in a rise in poverty levels and destitution. According to ancient sources, during that time

> ... wheat was scarce, the grains rotten and the food was little. People were stealing from others, babies weeping, young people wandering unemployed and the elders' hearts were full of sadness, their knees broken.[9]

When it became clear that the famine was coming to an end, the clergy invented another dream, in which Khnum—the God who protected the sources of Nile—promised that the suffering is coming to an end and fields will be flooded again. In exchange, the clergy demanded that the most fertile part of the land be dedicated to Khnum as being the messenger of good news. Naturally their demand was accepted, and masses of land were appropriated by the clergy—the God's representatives on Earth. After the famine, the clergy and Pharaoh had the coffers full and huge

---

[5] Geometry was also invented thanks to the Nile; each time the river flooded, the borders of the land were covered by water and had to be measured again.

[6] Sidney and Sylla, *op. cit.*, Chap. 2.

[7] An epic description of Joseph's role is given in the four-volume opus *Joseph and his Brethren* by Thomas Mann (NL in Literature 1929) (2005).

[8] The best narration of the story is provided by the Old Testament, Genesis.

[9] Quoted in Vandier, *La Famine dans l' Egypte Ancienne* (1979).

amounts of land in their property. However, state ownership of the agricultural economy also required central planning, which in turn had to be based on a large bureaucracy; as a result direct control by the Pharaohs weakened and economic power eventually came into the hands of the bureaucracy's most organized sector, i.e. the clergy. Bureaucratization had its own price, as the operation of the economy became less efficient and eventually declined.

3.3 Market Lessons from the Crisis: How Ancient Athens Avoided Famines

The famine in Egypt was a momentous event in antiquity and every other state thereafter hasted to ensure that it would be spared such a fate.[10] In ancient Athens, there were several periods of low grain production, but the city never suffered a famine.[11] That was thanks to its ingenious and comprehensive system of supply and regulation in the grain market ensuring that Athens was well-supplied in normal times and sufficiently so in times of hardship.[12] Part of the grain production took place in Attica, however most of Athens's supplies came from its colonies. The supplier states were geographically scattered, thus minimizing the risk of simultaneous destruction caused by an invasion or adverse weather conditions. The Athenian fleet protected trade routes from pirates and often transported the grain to Athens.

As in ancient Egypt, the grain was put by the merchants into large silos in the nearby port and then transported to Athens supplying the grain dealers who sold it to the final consumers. It is interesting to see how the market was organized and supervised. In order to avoid oligopolistic practices and tamper with prices, grain dealers were not allowed to transport more than 50 baskets of grain per day. It is noteworthy that, by doing this, Athens chose to regulate supply by imposing quantity quotas instead of price caps, as the latter often leads to hoarding and smuggling.

The implementation of rules was the duty of the grain guards who resembled what we would today call a market regulator.[13] Penalties for transgressors were severe and the public was anxious to see them punished. One should not believe, however, that practices always conformed to strict rules. Grain dealers often spread false rumors that crops had withered or that the fleet transporting it to Athens sunk,

[10] A description of such responses is given in Garnsey, *Famine and Food Supply in the Graeco-Roman World: Responses to Risk and Crisis* (1988).

[11] In O'Grada, *Famine: A short history* (2009, p. 14). Naturally, grain cultivation is facilitated by the temperate climate.

[12] One exception is the famine during the Peloponnesian War, which was caused by the embargo of Athens by Sparta. Since Athenians were forced to remain inside the Long Walls, famine was accompanied by diseases.

[13] "Their duties are, first, to see that the unprepared corn in the market is offered for sale at reasonable prices". Aristotle, *Athenian Constitution,* Part B, LI, I-4.

or that the merchandise was contaminated, in order to cause uncertainty regarding future sufficiency and thus fuel current demand and, by extension, prices.

Lysias—a famous orator and speech writer of antiquity—had been prosecuting a case in 386 BC against grain sellers arrested for concerted practices. In his speech at court, he reveals many malpractices by the grain dealers, and warns Athenians that:

> Seeing your misfortunes they get glad, so that they often hear of some before other people, and others they make up themselves; either that the ships in the Pontus sea have been destroyed, or captured by the Lacedaemonians, or that the market is closed, or that the peace treaties are about to be made void and wars will break out.[14]

In order to deal with abusive practices, the Athenian state doubled the number of grain guards and broadened their powers. Indeed the city had such a low esteem of grain dealers that, whereas in other areas of the market, it was the Parliament who appointed the auditors in charge of supervising transactions, the grain guards were appointed through public draw—obviously in a move to minimize the risk of unethical practices.

3.4 Economic Crises in Ancient Greece

3.4.1 Money Creation

In ancient Greece, the monetary system was metallic and the exchange value of each currency depended on the metal it was made of and its weight. Every city minted its own currency and, naturally, cities with silver or gold deposits found themselves having a great economic advantage. For example, Athens' apogee is closely related to the discovery of silver mines, while Philip II of Macedonia used the nearby goldmines to buy off lesser cities and be voted head of the League of Greece. When his son Alexander conquered Persia made immediate and extensive use of the seigniorage by issuing new money. It is said that when he arrived victorious at Susa, the first thing he did was to sit on the throne of Great Kings, and immediately order that the treasures of Persia are used to mint new coins so that he could finance further campaigns.[15] The *tetradrachm*, a coin depicting Alexander, circulated throughout his empire from India and Afghanistan to Egypt and Libya, was perhaps the first international currency in history, giving a sense of "monetary union" to the Hellenistic world.

[14] Lycias, *The Grain Dealers, 14.*

[15] See Gehrke, *Geschichte des Hellenismus* (2000, p. 43).

3.4.2 Money Corrosion

Besides mining, another–less costly–way to increase money supply was by scraping and collecting the filings from the coins and amalgamating them with cheaper metals to recast new ones with the same nominal value. Scraping was the first recorded method of tampering with the value of money and was a widespread practice, especially in periods when the state was in great need of further resources. Such a period was in the aftermath of the Peloponnesian War when Athens debased the currency to pay for the war obligations. The old coins vanished and the debasement was famously mocked by Aristophanes, foretelling Gresham's Law that "bad money drives away good money" as the latter is hoarded:

> We treat our best men the way we treat our mint
> The silver and the golden we were proud to invent
> These unalloyed genuine coins, no less,
> Ringing true and tested both abroad and [in] Greece
> And now they're not employed as if we were disgusted
> And want to use instead these shoddy coppers minted.[16]

The same danger was highlighted by Plato, who, in a more serious tone, urged that the quality of money in circulation should be monitored and protected. Plato also suggested that, in order to avoid both the risk of forgery and the imposition of heavy taxation when faced with urgent financing needs, Athens should temporarily repeal convertibility and issue fiat money.

3.4.3 Lending and Debt Crises

Interest-bearing lending was widespread; rich merchants were the usual lenders, but the same function was exercised by temples, which were collecting the donations of pilgrims or accepting their money for safekeeping.[17] Lending, as well as the exchange of foreign currency, was carried out over a small table (τράπεζα), which explains where the modern Greek name of financial institutions came from. The same was true of ancient Rome, where transactions also took place over a long bench called *banco* that gave rise to the modern term 'bank'. Bankers did not enjoy a good reputation in ancient Athens, and were frequently ridiculed by Aristophanes, while Aristotle condemns those who gain excessively by lending small amounts of money at high interest rates.[18] Isocrates attacked the greedy lenders in his *Trapeziticus*, an oration he wrote for a merchant who was taken to court by Passion, a *metic* banker, for not servicing a loan.

Episodes of bank crises were not uncommon in Athens as bankers—acting as financial intermediaries—were sometimes unable to return the deposits, even those

[16] Aristophanes, *Frogs*, 720.

[17] A detailed account is given by Millet, *Lending and Borrowing in Ancient Athens* (1991).

[18] Aristotle, *Nicomachean Ethics*, IV 30.

drawn from the temples. A major scandal erupted in the fourth century BC when deposits from the Temple of Athena given to bankers could not be redeemed and the vault of the Temple was set on fire to destroy records of the transactions.[19]

Unpaid debts could lead to the enslavement of the debtor, until Solon abolished debt bondage in the sixth century BC. Even so, however, difficulties and problems remained. Many Athenians were unable to pay back the loans they had obtained in order to purchase land and were in danger of losing it; without land possession, they would automatically be excluded from the city's public life. To avoid a political setback, Solon implemented the reform of '*seisachtheia*' (literally, shaking off debt burdens), a measure that erased the loans and allowed the farmers to keep the land and their political rights.

Solon's debt forgiveness was politically tarnished by what is perhaps the first recorded incident of insider information. Some of Solon's friends knew beforehand what would happen and rushed to get loans to buy land in expensive areas. After the law passed, they were also included in the debt forgiveness program, keeping the land and becoming wealthy overnight. The incident naturally caused a huge scandal and Solon himself was implicated by his political adversaries. According to Aristotle,[20] the hostile reactions against the abuses of debt forgiveness policy were one of the reasons that Solon spent many years in self-exile to Egypt.

When sovereign debt reached non-viable levels, ancient rulers would take measures that in essence are not dissimilar to the policy responses during later crises. Unable to meet his obligations, Dionysius of Syracuse used a drastic method to pay back the loans he had taken from citizens, without even bothering with scrapping the coins.[21] He simply issued a decree that all money in circulation had to be surrendered to him and then stamped the coins with twice the original remuneration and gave half of them back, saying that they could be used at the new nominal value. The other half was used to pay off his debts, effectively imposing a 50 % cut at his subjects' real balances and purchasing power.

3.4.4 Taxation and Representation

When Pericles started preparations for the Peloponnesian War, he imposed a property tax to fund the construction of naval vessels, and the episode evolved as the first institutional trade-off between taxation and political representation.[22] Athenians reacted strongly and consented only when Pericles promised to broaden

---

[19] For a description see Cohen, *Athenian Economy and Society* (1992).

[20] 'People who knew him distanced themselves because of his debt cutting policy'; Aristotle, *Athenian Constitution*, XI.

[21] This incident is cited by Reinhart and Rogoff, *This time is different: Eight centuries of Financial Folly* (2009, p. 174).

[22] The incident is mentioned as the first case of connecting taxation and representation by Ferguson, *The Cash Nexus: Money and Politics in Modern History 1700–2000* (2001, p. 83).

their participation in the allocation of state revenues. The Parliament was endowed with new powers with regard to state expenditure and by using its new-found influence Pericles attempted to take control of all political power in the city.[23]

In general, Pericles was known more for his tendency for public spending rather than his thriftiness. For example, in order to undercut the power of the Supreme Court (*Areios Pagos*), he instituted allowances for members of public offices to facilitate participation of poorer citizens. The measure drew a lot of criticism such as for burdening the budget and degrading the function of representative institutions.[24] In order to fund his grandiose projects, Pericles appropriated the treasure of the Delian League—i.e. the gold deposited by friendly cities for emergency cases—thus angering the Athenian Alliance. Athens's reputation as a trustworthy metropolis was severely tarnished leading to the weakening of its rule and the final defeat by Sparta in the Peloponnesian War. The city then entered a long phase of decline, from which it would never fully recover.

3.5 The Evolution of Economic Thinking

After the defeat of Athens and the ensuing crisis and decline, the main essays and analyses on economics were written as a response on how to reorganize the city and enable it to recover power and influence. The main contributions came from Plato, Xenophon and Aristotle, and the key principles of their economic thought are discussed below.

3.5.1 Plato's Economic Advice

Plato (427–347 BC) immediately realized the need to reorganize Athens and made a full-fledged proposal for the ideal *Republic*, whereas several economic prescriptions are laid out. He recognized that redistribution and fairness is the key for the city to recover and proposed specific limits for restraining inequality; specifically that the highest cohort of wealthy citizens is not endowed with more than five times the value owned by the lowest.[25]

In matters of economic organization, Plato argued in favor of labor division based primarily on social status rather than on skills specialization. Since Plato was deeply conservative, his argument does not stem from any search for economic efficiency—as was the case with Xenophon—but rather from contempt for those

[23] In the words of Aristotle, this "...resulted in emboldening the multitude, who brought all the government more into their own hands". *Athenian Constitution*, XXVII.

[24] "Some critics accuse him of thereby causing a deterioration in the character of the officers"; as noted by Aristotle, *op.cit.*, XXVII, 4.

[25] Plato, *Laws*. The ratio maximum/minimum = 5 is reminiscing of the Pareto 80/20 law on income distribution; see Sect. 13.1.4.

practicing commerce and other lay vocations. Public life, according to Plato, should be a privilege accorded only to those who do not need to work for a living and explicitly argues in favor of the exclusion from politics of everyone who has to work to make ends meet. They were not fit to rule the city because

> freemen who of their own will become the servants of the other classes in a State, and those who exchange and invoice the products of husbandry and the other vacation, some sitting in the market-place, others going from city to city by land or sea and giving money in exchange for money or for other productions—the money-changer, the merchant, the ship-owner, the retailer, should not put in any claim to statehood or politics.[26]

3.5.2 Xenophon's Economic Organization

Xenophon (430–354 BC) was a wealthy Athenian with a large acreage of land and his preoccupation was about finding the best way to utilize it. His treatise "*Oeconomicus*" literally refers to house management concerning agriculture and production, and thus he became the godfather of the subject matter. Xenophon wrote extensively on the division of labour, the exchange of goods as well as some ideas on the comparative advantage in trade. Of all his writings, those most related to the re-organization of Athens were concerned with the increase of state revenues as a critical condition for financing the recovery.

In his work *Ways and Means*, Xenophon describes how the city could exploit the mines in order to promote production and also increase state revenues. He suggests a system of labor division, which allocates duties and types of employment (military, merchants, etc.) among various categories of citizens in order to raise efficiency and increase security for the city. One of the most original ideas in this book is Xenophon's proposal that the state participates in the operation jointly with the private sector, in order to spread entrepreneurial risk.[27] Anticipating the strong reaction from both the proponents of privatization and the supporters of state ownership of his time, Xenophon reassures them that such a public-private partnership would not be detrimental but rather beneficial to both parties:

> Nor need you apprehend, sirs that a state mining company, established on this principle, will prove a thorn in the side of the private owner, or the private owner proves injurious to the state. But rather like allies... render each other stronger....[28]

This is an early formulation of views still voiced today with almost identical arguments. Yet, despite the insightful recording of economic phenomena of his era, overall Xenophon was not able to structure his approaches into a cohesive theory. It

---

[26] Plato, *Statesman*.

[27] "[I]t is possible, of course, for private persons to unite in the same way, share their fortunes and minimize their risks". Xenophon, *Ways and Means*, 31. His plea is actually addressed to the ten Tribes which owned the mines.

[28] Xenophon, *op.cit., 32.*

was Aristotle who put practices and policies into a coherent philosophical system and influenced economic thinking for centuries to come.

3.5.3 Aristotle's Economic Philosophy

Aristotle (384–322 BC) was a great polymath who investigated, examined and wrote on almost all fields of life and science: from physics and biology to law and astronomy, submitting views on such diverse issues as the optimal manner to distribute goods, the use of money or how to desalinize sea water. Aristotle established an anthropocentric philosophy, based on society's needs and not on religious doctrines as was the established tendency at that time.

Whereas Xenophon only described economic phenomena without attempting to explain them, Aristotle was the first to conceive economic thinking as a normative theory, connecting it with the rest of his philosophy in *Nicomachean Ethics*.[29] In the fourth century BC, Athens was still impacted by its defeat in the Peloponnesian War and, at the same time, feeling the pressure from the increasing power of Macedonia, Aristotle's place of origin. In order for Athens to get back on its feet and Greece to avoid subjugation to the Persians, Aristotle established a theory of social organization and individual behavior which would revive the power of the city and the wellbeing of its people.[30] His major contributions are the following:

Real and Financial Economy He observed that, besides production, there are also other activities and phenomena, including wealth that can be loaned to someone else and conspicuous consumption that does not cover real needs. He therefore distinguished economic activity between two spheres, the "*oeconomic*" one, which concerns production and management of essential goods, and the "*chrimatiki*" (i.e. financial) which concerns profit and rules of exchange. He was among the first to discern the role of money both as a medium of exchange and as storing value.

Utility of Consumption Aristotle calls knowing what is in one's best interest a key form of knowledge, and prudent the person who knows what concerns him and tries to achieve it.[31] He then formulated the concept of diminishing marginal utility,

[29] Nicomachus was the name of both his father and his son, and this fact gives his work a notion of moral accountability to the older and of teaching the younger generations. By historical coincidence, Adam Smith, the founder of modern economic theory, also started out as a moral philosopher.

[30] In other areas Aristotle was conforming to the limitations of his time. For example he was a defender of slavery, which he considered a natural phenomenon, and even suggested specific measures to make managing slavery more efficient.

[31] Aristotle, *Nicomachean Ethics* (VII, 30): "For people seek their own good, and suppose that it is right to do so. Hence this belief has caused the word 'prudent' to mean those who are wise in their own interest...".

according to which if more is consumed of a product, additional pleasure diminishes. In the same spirit, Aristotle wrote that there should be limits to wealth accumulation. Beyond those limits, wealth loses its usefulness and becomes dangerous, essentially describing a theory of diminishing returns. Aristotle was also the first to philosophically establish the concept of intergenerational solidarity, arguing that the primary task of parents is to ensure the wellbeing of their children and focus on their upbringing, while children in turn should take care of their parents when they grow old.

The Median Aristotle's greatest contribution to economics is the establishment of the "median" (μέσος) as a concept that should guide people's choices, actions, relationships and their way of life. Essentially, it is a description of the search for the optimal result, which is achieved after specific effort is employed—no more, no less. Otherwise, "temperance and courage are equally destroyed by excess or deficiency, but are preserved by the median."[32]

Mathematical Rules Aristotle was in close contact with the Pythagoreans whom he studied and, by using their mathematical findings, attempted to quantify his economic theories of the mean. When he analyzed the three concepts of economic justice to describe the fundamental economic functions, he suggested at the same time quantitative rules for how they could be actually implemented:

Employing the concept of *compensating fairness*, an injustice should be removed in a way that results in *ex post* equality for both parties; this equality is calculated by using the arithmetic mean, which automatically removes the initial deviation ("*for it exceeds and is exceeded by an equal amount; this is intermediate according to arithmetical proportion*").[33] The simple arithmetic mean is not an absolute concept, however, and when more individuals are concerned, remuneration should be calculated with regard to the status and other extraneous characteristics of each one ("*the intermediate not in object, but in relation to ourselves*"), suggesting a kind of a weighted mean.[34]

In the case of *distributive fairness*, he argues that the distribution of goods should take place according to the amount of effort spared by each member as compared to the whole. This notion refers to the geometric mean.[35] Finally, in the case of

[32] Aristotle, *op.cit.* (V, 25).

[33] Aristotle, *op. cit.* (V, 35).

[34] Aristotle, *op. cit.* (V, 5).

[35] An analogy in modern economic theory is a Cobb-Douglas production function, where output is the weighted geometric mean of input factors and each of them is rewarded proportionally to its intensity coefficient.

exchange fairness, when two parties want to engage in a transaction but diverge on setting the price, Aristotle suggests that they compromise equi-proportionally. This is equivalent to agreeing on the harmonic mean of two prices.[36]

As ancient Athens remained in decline, other views were also expressed regarding economic behavior and the pursuit of wellbeing. The stoics and the so called Cynic philosophers propagated the benefits of a frugal life by adhering to the means and rules of nature. In contrast, Epicurus argued[37] that life can only be happy insofar as it is prudent, moral and just, and vice versa; the latter implying that a person who does not possess the essential goods cannot live happily. Thus the question arose on how to identify individual interest, find the best means for serving it, and finally measure it. According to Aristippus, these properties are impossible to be quantified as a person's enjoyment depends on many factors, both subjective and economic.[38] All the above arguments would strongly resonate in the theories of utilitarianism in the nineteenth century.

3.6 Modes of Production and Economic Thinking in the Ancient East

The emergence of economic thinking as a result of economic problems and crises was not unique in the Mediterranean world. Other civilizations were also challenged on how to deal with population subsistence, but the organizational structures as well as the outcomes were not the same in all cases. In the empires of Asia there was widespread use of central economic planning in order to deal with large scale production and distribution.

3.6.1 Eastern Empires

In ancient India, economic rules seem to be quite detailed and widespread; for example, to ensure the normal function of credit the servicing of debt was guaranteed by passing the obligation to the debtor's children as modern overlapping-generations models would have assumed by the intertemporal budget constraint. There were also functions of central planning to coordinate production and distribution of food. In ancient China, the best known thinker was Fan Li who

---

[36] Suppose that one sells a good at a price P1 and another is offering P2 to buy it. If $P1 < P2$, transaction is easy. But if $P1 > P2$, there is no agreement. In which price P* is fair the transaction to take place? If both parties concede by the same proportion (say λ), the fair price is found by setting $P^* = (1 - \lambda)P1 = (1 + \lambda)P2$; eliminating ($\lambda$), P* is easily obtained as the harmonic mean.

[37] Epicurus, *Principal Doctrines*, (I. Ethics, V): "It is neither possible to live joyously without also living wisely and beautifully and rightly nor to live wisely and beautifully and rightly without living joyously; and whoever lacks this cannot live joyously".

[38] Kanellopoulos, *Modern economic thoughts of ancient Greeks* (1996, p. 365).

lived in the fifth century BC and is considered as the world's first economist.[39] He studied the fluctuation of prices due to changes in demand or seasonality. In order to avoid undesirable price variations, he suggested that the state intervenes in product markets during periods of strong growth or pronounced contraction, so that the price remains as stable as possible. In essence, this was a countercyclical price stabilization policy, with the state buying goods when prices fell and selling them when prices increased.

Two streams of economic thinking emerged in China and remain in perpetual competition ever since: one based on the teaching of Confucius proposing a moral paradigm of man, and another based on the needs and interests of the state and its empowerment. According to Confucianism, the empire must not interfere in economic matters; agricultural production should be decentralized and organized according to tradition, while trade is kept limited as a parasitic activity. The supreme target of governance should be the well-being of ordinary people, which is achieved when their basic needs—and not beyond—are met. Confucius gives emphasis to the moral basis of governance as a fundamental prerequisite to secure the trust of its subjects. Only then are people willing to be lead by decree and conform to punishment without shame. Similar views would be expressed by Locke in seventeenth century England. Confucianism took a stance in favor of the traditional way of life, frugality and measure in everything. It urged avoidance of extremes, and considered it reproachful for someone to be rich and noble in times of crisis; but equal scorn is accorded to anyone being poor during the good times.[40]

The mandarins represented Nomocracy, the rule of Law, and were proponents of reforms in the economy in order to expand trade. They believed that economic rules should not stay unaltered; rather, they should adjust according to the needs and the prevailing conditions. The mandarins established state monopolies in salt, iron, tea and wine, and by subsequently imposing tariffs, increased the imperial treasury; the increased revenue was then used to finance the military (and themselves, naturally). By abolishing the right of communities and merchants to mint their own currency, they granted this right exclusively to the emperor, something that obviously increased even more the latter's revenue and power. The conflicting views between Confucianism, expressing the traditional sections of the Chinese countryside, and Nomocracy, which expressed the interests of commerce, are laboriously and vividly recorded in the book *Salt and Iron*, as minutes in a meeting of representatives of both groups in 81 BC.[41]

---

[39] In a survey published by the English newspaper Telegraph in 2011, Fan Li tops the list with the ten most famous economists of all times (3/7/2011). Nationality of voters was not reported.

[40] Confucius, *Analects*, 16 (1979).

[41] The author of the book is unknown.

3.6.2 Modes of Production

Broadly speaking, two different modes of economic organization had emerged in ancient times: one dominant in the Mediterranean (ancient Greece and Rome) and another in the vast empires of Egypt, Persia and the East. In brief, the major characteristics of the two modes are the following:

Ancient Mode of Production The city (*polis*) is the epicenter, while the surrounding area is used as farmland. The cultivation of private lands, both small and large, is crucially dependent on slave ownership; slaves farm the lands and undertake all laborious work involved in mining, sea transportation etc. Manufacturing activities are of small-scale and mainly include pottery, ship building, metallurgy, furniture and weaving. Trade is particularly developed between cities and states.

Asian Mode of Production The state is the great landowner, while the farmers are usually landless or own a very small amount of property. Agricultural production is essentially under state command, and the state is also responsible for distributing the surplus and building infrastructure. The countryside is densely populated and segregated from city life and trade is limited. In these states, recording systems are developed to assist production and distribution planning. For example in ancient Persia, a system of taxation and a detailed auditing of state expenditure were set up, surpassing any contemporary analogue.[42] Wittvogel (1957) calls these "hydraulic states", because they are organized in order to manage the large scale irrigation infrastructure that is necessary to cultivate the land. Though similar regimes ruled ancient India and China, perhaps the most representative application of this model took place in Egypt.

Together with the later system of feudalism, these ancient forms are known as "pre-capitalist modes of production" and were extensively studied by economists and historians, particularly of a Marxian tendency. One issue that concerned those theorists was whether it was inevitable that all economies would have to pass through every stage, each one leading inexorably to the next, or one could intervene and alter the course of history. The answers given to this question were boisterous.

In antiquity, Aristotle was the one who had made the distinction between two types of societies and he coined the term "despotism" to describe the organization of the Persian state after a visit he made there. Later, German philosopher Hegel would develop the theory of the separate, distinct stages of history: the slave economy in antiquity, the system of feudalism in the middle Ages, and the age of constitutionalism in modern times.[43] Hegel attempted to distinguish them based on

[42] Darius set up twenty regional areas, the *satrapies*, so that each nation was assessed for taxes; Herodotus, *Histories*, Book 3, 90–94.

[43] For a detailed description of these phases see: Hegel, *Philosophy of History* (1956).

structural features, but later he also classified them geographically by attributing a speedier transition into the next stage to Europe as compared to the sluggishness of the Eastern world.

In the Marxist theory, the moving forces are no longer sets of ideas and beliefs, but class economic relations. Marx was the first to identify that the criterion distinguishing the modes of production is not geography, but the organization of labor, the means of production and the social relations of ownership that perpetuate them.[44] In essence, however, the structure of historical evolution remains along Hegelian lines. Marx's classification and analysis of the pre-capitalist modes of production during antiquity closely echoes Hegel's by distinguishing them between the ancient slave-owning model and Asian despotism.

References

Cohen E (1992) Athenian economy and society. Princeton University Press, Princeton

Confucius (1979) The analects. Penguin, London

Ferguson N (2001) The cash nexus: money and politics in modern history 1700–2000. Penguin, London

Garnsey P (1988) Famine and food supply in the Graeco-Roman world: responses to risk and crisis. Cambridge University Press, Cambridge

Gehrke HJ (2000) Geschichte des Hellenismus. Oldenburg, Munich

Hegel G (1956) The philosophy of history. Dover, New York

Heilbroner R, Milberg W (2008) The making of economic society. Pearson Education, Upper Saddle River, NJ

Hicks J (1969) A theory of economic history. Clarendon, Bristol

Kanellopoulos A (1996) Modern economic thoughts of ancient Greeks. Livanis Publications, Athens (in Greek)

Mann T (2005) Joseph and his Brethern (trans: Woods JE). Everyman's Library, London

Marx K (1971) A contribution to the critique of political economy. Lawrence & Wishart, London

Millet P (1991) Lending and borrowing in ancient Athens. Cambridge University Press, Cambridge

O'Grada C (2009) Famine: a short history. Princeton University Press, Princeton

Reinhart C, Rogoff K (2009) This time is different: eight centuries of financial folly. Princeton University Press, Princeton

Sidney H, Sylla R (2005) A history of interest rates. Wiley, New York

Vandier J (1979) La Famine dans l'Egypte Ancienne. Arno Press, New York

Wittvogel KA (1957) Oriental despotism; a comparative study of total power. Yale University Press, New Haven

[44] Marx, *Contribution to the critique of political economy* (1971).

4 Economic Crises and Practices in the Roman and Byzantine Era

Abstract

The Roman Empire was too much relying on slave labour to feel the need for technical innovation, but nevertheless the recurrence of economic crises exercised a constant pressure to improve policies and institutions. When a real-estate bubble collapsed, Caesar invented a scheme of financial support to save households and banks. After the monetary collapse in the times of Diocletian, the introduction of coinage based on Gold restored confidence in the Eastern Roman Empire and ushered in a long period of flourishing trade and economic reforms. Economic thinking was advanced too, ranging from pragmatic rules on lending and profit to the concept of single tax and distribution theories.

4.1 Expansion Without Innovation

Next historical era is the Roman Empire. In comparison to ancient Greece, the scale of economic activity went up by a multiple but the freedom of production factors lagged behind and was even curtailed. The majority of the labor force consisted of slaves, with extensive slave ownership covering the labour needs in the large Roman land estates (*latifundia*). The slave trade became a multinational enterprise, which thrived after the Roman conquests in Africa and Asia; in fact sometimes it was the main objective of the campaign.

It seems that extensive slave ownership was the major reason that Rome—as well as Athens and Sparta previously[1]—showed no interest in developing and introducing technological innovations in production techniques, despite advanced

[1] Montesquieu notes that "Romans considered commercial and technical activities as suitable only for slaves", in line with the views of Plato previously discusses in Sect. 3.5.1. Montesquieu, *Considérations sur les causes de la grandeur des Romains et de leur décadence* (1968, Chap. 10).

N. Christodoulakis, *How Crises Shaped Economic Ideas and Policies*,
DOI 10.1007/978-3-319-16871-5_4

technical capabilities of the Empire in other sectors. Rome, for example, promoted military technology, invented multistoried buildings and constructed imposing infrastructures such as aqueducts, sewage systems, and international road links. But not machines. In a characteristic episode, Greek engineer and mathematician Hero of Alexandria in 150 BC had invented a device resembling the steam engine. Hero's engine was a rocket style jet engine which spins when heated. Since the motive to reduce costs did not exist in the Empire, Hero could only showcase his invention in religious festivals.

Without pressure from production, innovation cannot thrive and further build upon existing scientific knowledge, as copious examples from history repeatedly demonstrated.[2] It is said that when an inventor presented Vespasian with the design of a water mill, the Emperor forbade its construction so that "the masses can continue to earn a living".[3]

Perhaps the same reason explains why, despite the long life of the Empire, hardly any noteworthy theoretical contribution in the economics of production took place. Rome seemed to be content to let the production system relying on slavery and concentrated on devising rules and institutions to protect ownership rights and facilitate commercial activity in its vast Empire. As domestic political stability was an essential pillar of outward expansion, the Emperors often had to intervene in order to alleviate various crises occurring either in real-estate or the supply chain of Rome.

4.2 Crises in the Roman Empire

4.2.1 Lending and Housing Crises

Lending money at interest was allowed and it was not unusual to see debtors unable to meet their obligations. One such occasion was the real-estate crisis that hit Rome after the Civil War, the first recorded manifestation of the bubble. According to historian Suetonius, many Romans who had borrowed large amounts to buy residencies in Rome were now begging for debt-forgiveness as plummeting asset values resulted in their debts becoming burdensome. Caesar, fearing that a debt relief may cause banks to collapse, decided to directly subsidize debtors by an amount of money to the tune of 25 % of total debt to pay their lenders.[4]

[2] The steam engine was rediscovered and finally put in production in the course of the Industrial Revolution, more than eighteen centuries later.

[3] See Cipolla, *Before the Industrial Revolution: European Society and Economy, 1000–1700* (1994).

[4] Suetonius, *The Twelve Caesars*, Book I: XLII. The US government did the opposite after the 2008 subprime mortgage crisis; by providing direct support to the banks made depressed households vulnerable to the pressures of lenders. In most cases, lenders foreclosed the mortgaged properties.

Households' debts were further relieved by the easing of monetary policy under Augustus. When he "brought the treasure of Ptolemies into Rome at his Alexandrian triumph, so much cash passed into private hands, that the interest rate on loans dropped sharply, while real estate values soared".[5] To further support the real estate market, Augustus would grant interest-free loans for fixed periods to anyone who could offer guarantees for twice the amount.

4.2.2 Famines and Government Policy

Noteworthy is the policy response by Augustus when a famine struck the capital in year 6 AD. Rome had set up public commissions to organize the transportation and distribution of grain, while also aiding the poor by handing out grain for free. When grain production shrunk due to prolonged periods of infertility, Augustus interfered in both sides of supply and demand: while paying more attention to the needs and interests of farmers, he also decreased the number of free beneficiaries in the city.[6] To this effect, he expelled all members of gladiatorial schools, foreign residents—with the exception of physicians and teachers—as well as several of household slaves.

After the situation got stabilized, Augustus decided "to discontinue the free supply of grain to the citizens, reliance on which had discouraged Italian agriculture". But he finally did not go ahead and the reason is also revealing: Suetonius says that Augustus refrained because sooner or later "... some politician would be obliged one day to revive the dole as a means of ingratiating himself with the people"; a clear manifestation of massive pork-barrel politics being at work in ancient Rome.[7] As a matter of fact cycles of spending sprees and contraction became the routine in the endless chain of imperial succession. Each new Emperor rushed to increase army salaries and distribute benefits to the people, before embarking on new efforts to enhance revenues from further taxation, property confiscations or new campaigns.

Eventually these methods were exhausted and authorities had to devise new instruments to finance their plans. Two Emperors are worth mentioning for their inventiveness: Nero first imposed a lump-sum contribution with payments required in precious coins and then debased the currency by about 10 %, initiating a policy that was eagerly followed by his successors.[8] At the beginning of third century the currency had lost about half of its value and Caracalla promulgated a new currency. But as expenditures could not be contained and debasement was not as effective as before, he sought more innovative ways. When his Edict in 212 granted Roman

[5] Suetonius, *op.cit.*, Book II, XLI.

[6] Foretelling similar policy prescriptions suggested by Malthus in the nineteenth century AD; see Chap. 6.

[7] Suetonius, *op. cit.*, Book II, XLII.

[8] Suetonius, *op. cit.*, Book VI, XLIV.

citizenship to ex-slaves, Caracalla seized the opportunity and doubled the inheritance tax they would be liable to pay so that imperial revenues could be further maximized.[9] The Edict left a mark on history and was copied several times afterwards. A modern inspiration can be found in the way that some European states grant investment-based citizenship to foreigners, (e.g. Austria, Cyprus, Malta, etc.).

4.2.3 The Crisis Under Diocletian

During the late period of the Empire, several economic crises were erupted and precipitated its demise. In the late third century, Diocletian tried to secure its power, challenged by barbarian raids, and to reinstate the Roman way of life, threatened by the advent of Christianity. Both required substantial financial means, to pay for the army fighting the raiders and to organize magnificent ceremonies at the temples of the ancient gods to overshadow the glare of new religion. But state coffers were empty, either because usurpers had depleted gold reserves from the treasury, or revenues were dwindling as commerce was severely hit by the raids.

In an effort to avoid bankruptcy, Diocletian devised a balanced-budget rule by flexibly adjusting the tax rate so as to meet expenditures.[10] But the constant rise of state payments meant that the rule squeezed communities to a point of revolt, and the Emperor soon resorted to the traditional policy of currency debasement. As adulterated coins were put in circulation with the same nominal value, prices went up as predicted by the quantity theory of money, and a vicious cycle of inflation and hoarding followed suit. To control galloping inflation, Diocletian issued the *Edict on Maximum Prices*, one of the most far-reaching price regulations to be recorded in history.[11] Again, the result of this measure was predictable: commodities priced above the cap simply disappeared from normal transactions and traded in the black market; at the same time, commodities priced below the ceiling skyrocketed, hitting purchasing power. Produce was hoarded by farmers, the grain trade collapsed and famine was taking its toll.[12]

To pay his starving armies, Diocletian further increased the circulation of debased coinage and pressed for higher tax revenues, aggravating the problems at haphazard levels. The devastation of people was so deep that invasions were deemed as a lesser evil or even a salvation from further impoverishment, and whole regions exasperated "as a result of this exaction of taxes… sought the help

[9] Described in Tainter, *The collapse of complex societies* (1988, p. 137).

[10] Described in Tainter, *op. cit.*, p. 143.

[11] A description of the Edict is provided by Doyle (1976).

[12] For a detailed record of famines and pestilences in the late Roman Empire, see Stathakopoulos, *Famine and Pestilence in the late Roman and early Byzantine Empire: systematic survey of subsistence crises and epidemics* (2004).

of the barbarians".[13] Monetary collapse and over-taxation coincided with the disintegration of the Empire and sped up its ultimate fall.

4.3 Currency Reform in Byzantium

When Constantine became emperor of the Eastern Roman Empire, building Constantinople as the new capital, he looked for a new monetary system that could stand the test of time and minted the golden solidus. To ensure the required gold reserves, he forbade private gold holdings, all possible sources of gold were amassed by the state and a senior official was appointed as responsible for the empire's mints. The issuing office was named *Sacrae Largitiones*, a kind of central bank of the time.[14] After Christianity was established as the official religion of the empire in 380, Emperor Theodosius quickly outlawed ancient temples and appropriated their gold reserves to further strengthen the new monetary regime.[15]

The purity of the coinage made it reliable for trade transactions and established it as a reserve currency throughout the known world. Price levels rapidly fell and the Eastern Empire underwent a long period of remarkable economic stability, while Rome was being destroyed by raids and the West entered the dark period. The solidus remained in circulation for at least 800 years, until the mid-thirteenth century, becoming the longest lasting and most stable currency in the history of mankind.[16]

Constantine's monetary reform was at a later stage complemented by the lawmaking initiatives of Justinian, who issued the Codex, a compilation of imperial enactments regarding economic transactions.[17] The Codex introduced the notion of "fair trade", according to which the value of a commodity is the price obtainable in the open market, and also laid that an acceptable profit margin would be 10 %.

Though the solidus was retained as the basic currency, authorities also minted lower-value coins for everyday transactions which were not convertible into gold. These coins often came in and out of circulation for two reasons: first to prevent counterfeiting and, second, to take advantage of the fact that not all of the withdrawn coins were actually returned and exchanged for new ones, thus surrendering some value to the state.[18] As in other places, the most widespread form of currency

[13] Quoted by Tainter, *op. cit.*, p. 147.

[14] Papasotoriou, *Byzantine high strategy: 6th–11th centuries* (2001, p. 55).

[15] The destruction of the ancient temples covered other needs as well. Most of the building blocks used for the erection of the Church of Hagia Sofia came from ancient temples, with leftovers used for the construction of Constantinople's aqueduct.

[16] The English golden sovereign has been in circulation since 1692, so far less than half the duration of the solidus.

[17] For a non-legalistic description of the content and the effect of the codex see Rosen, *Justinian's Flea: The First Great Plague and the End of the Roman Empire* (2008).

[18] As described by Harvey, "in both cases it is almost certain that the transaction was at the cost of the producer", *Economic expansion in the Byzantine Empire 900–1200* (1997, p. 138).

misappropriation was coin-clipping and the illegal minting of new coins. To discourage such practices, Byzantine coins featured the busts of emperors on one side and of saints on the other; implying that any attempt to counterfeit them would be accompanied by harsh punishment, imposed either by a worldly or transcendental power.[19]

This practice, however, did have its side-effects. In 692, when Justinian II minted coins depicting Jesus Christ, the Muslim merchants found an excuse in the Quran's prohibition of religious images to denounce Byzantine coins and mint their own. Al-Malik (646–705) instrumented an extensive monetary reform, which unified the circulation of imageless coins throughout the Arab-controlled territories, and put an end to the co-existence of the Byzantine solidus with local currencies. This monetary unification played a crucial role in the development of trade in the Islamic world and its subsequent economic advance.

4.4 Lending and Taxation

Another issue the Byzantine[20] economy dealt with was lending at interest. In its attempt to appeal to the poor, Christianity had from early on condemned the greed of the rich and adopted a hostile stance against usury.[21] In the fourth century, Church pundits popularized the principles of the new religion, highlighting debt forgiveness as a God-pleasing act, which can eventually lead to absolution ("as we forgive those who trespass against us"). Saint Basil the Great argued that charity is an obligation of the rich to God, while his younger brother, Gregory of Nyssa, wrote *Contra Usurarios*, with warnings directed to those lending money at interest as well as to those who "recklessly borrow and unwittingly fall in the trap of interest".[22]

Contrary to the prohibitions enforced in medieval Western Europe, lending at interest was acceptable in the Eastern part, as long as the rate was not excessive; only the Church was entitled to interest-free loans from merchants. Interest rates for the laymen were regulated by the state and higher ones were allowed if a risky activity was to be financed. For example the 'sea-loans' given to merchant ships were wiped out in cases of shipwreck or pirate raids. Pricing the risk, interest rates for sea-loans circa 800 rose from 12 to 16 %, to take account of the increased pirate activity in the Mediterranean.

[19] In Europe, coin-clipping was eliminated in the sixteenth century with the introduction of indentations on coins.

[20] The term Byzantine came in use as a synonym of the Eastern Roman Empire only after its fall, but it has since prevailed and I use it in this book as an equivalent alternative name.

[21] In His Sermon on the Mount, Jesus Christ advises the rich who want to be virtuous to give interest-free loans, because only sinners lend money at interest: "...And if you lend to those from whom you hope to receive, what credit is that to you? Even sinners lend to sinners, to receive as much again. But love your enemies, and do good, and lend, expecting nothing in return" (New Testament, Luke, 6:34–36).

[22] St Gregory of Nyssa & St Basil the Great, *The rich, the poor and usurers*.

The Empire also paid great attention to revenue collection and fiscal management. The bulk of state income was used to pay for the army and public administration; any remaining surplus was spent on public works or for distributing bread for free to the city poor, in keeping with Roman tradition. Taxes on farmers were paid in kind and rates varied according to production levels. Merchants paid their taxes in gold or silver (the so called *chrysargyron* tax), and tariffs were imposed on imported commodities. The tax regime was adjusted, however, following the state of the economy and responding to the pressure exercised by various groups; for example the chrysargyron tax was abolished by Emperor Anastasius, who wanted to bolster commerce. Sometimes farmers paid taxes based on acreage rather than production levels, while at times a "village tax" would be imposed on an entire community, in addition to the latter's obligation to sustain a specific number of soldiers.[23]

Eventually taxation became so harsh and extensive that several tax revolts broke out. The most threatening took place in Thessalonica in 1341, where merchants and seamen deplored excessive taxation and rising interest payments, and managed to hold the city for 6 years.[24] But there was a shift in economic counseling as well, best summarized by Gemistus Pletho (1355–1450), a polymath in state affairs and Greek philosophy. Gemistus understood the exasperation and resentment caused by the existence of a multitude of different taxes and argued in favor of a single-tax system. Moreover, he proposed a fair distribution of income among what he identified as the three constituent parts of society: farmers, artisans-merchants and the state.[25] But this economic rationalization came too late to reverse economic decline and social disorder, and the Empire finally fell to the advancing Ottomans in 1453.

4.5 Byzantium's Pioneering Economic Thought

The economic approaches that were developed in Byzantium seem to be more sophisticated and functional compared to the ones prevailed in Western Europe during Middle Age. Several of them are also found to be compatible with later economic theories and some present day approaches, as for example:

The theory of value: In Byzantium, the prevailing view held that value "arises" from the market exchange based on the relation between supply and demand; in the West, theologians devoted enormous mental efforts pondering on the "fair price"

[23] Similar practices applied to the agricultural communities in Germany during Middle Ages, and this is perhaps what led Marx to find analogies between the Byzantine system and the German mode of production. Actually, he briefly refers to the Slavonic mode of production which prevailed in the Balkans and was based on small parcel land ownership.

[24] For a detailed account, see Kordatos, *op.cit.* (1928).

[25] An account of the economic views developed by Gemistus is given in Spentzas, *Philosopher of Mystra* (1996).

before the concept was finally denounced by Thomas Aquinas in the thirteenth century.

Cost of money: In Byzantium, lending was regulated by a series of legal provisions and the mechanism of interest rate was acceptable; in the West, Christians were forbidden to lend at interest.

Commerce and profit: In Medieval Europe, the Church stigmatized both activities as unethical; in Byzantium, commerce and profit were admitted as fully compatible with Christian teaching.

It is true that these approaches were not part of any comprehensive economic theory but rather a mosaic of traditions and practical compromises. Nevertheless, they supported an economic process far more flexible than the static state characterizing the Western part of Europe for many centuries. Some factors that explain this advancement are the following:

4.5.1 Institutional Continuity

The Byzantine Empire was established as the institutional successor of the Roman one, thus it was natural to maintain the social fabric and assimilate the higher echelons of the urban population in state administration. As early as in the sixth century, Justinian codified and enriched the Roman law and introduced rules regulating profit, commerce and lending. In contrast, the Roman political structure vanished in Western Europe under the new conquerors. Economic activity was interpreted and legitimized according to a strict set of rules stipulated by the Holy See.

Justinian's modernization is recognized, among others, by leading Marxists. According to Antonio Gramsci, the ruling class of the Eastern Empire was the one which demanded and succeeded in having a codification of the Roman law, so that a concise and universally applicable legal framework would substitute local and customary law, as the latter was often arbitrarily governed by the whim of powerful individuals or state officials.[26] Gramsci went as far as to claim this measure as a precursor of the "constitutionalism movement" that emerged in seventeenth century Western Europe.

4.5.2 Commerce

Byzantium's geographic position covered the Roman Empire's main commercial roads from Mesopotamia to the Mediterranean and from Egypt to the Danube. Therefore, strong ties were forged between the merchants, who wanted to safeguard trade roots and communications to boost their activities, and imperial authorities who needed the tariff revenues to pay for the army and ensure the adequate supply

[26] See Gramsci, *Prison Notebooks* (1975, p. 47).

of its cities.[27] In Western Europe, travel remained an extremely dangerous affair for many centuries and trade hardly existed as a long-distance activity.

4.5.3 More Freedom in Production

Arable lands retained their centuries-old structure and ownership status: small parcels of land, stable taxation and community-based production. This structure discouraged the emergence of large estates and motivated farmers to expand cultivation and raise output. Even dependent peasants (*paroikoi*, i.e. by the house) were not slaves in the traditional sense, but rather "half-free".[28] They had no right to leave the land, but their master had no right to dislocate them either. In contrast, the economy in the West took a dive after the fall of the Western Roman Empire. Adam Smith notes that "it seldom happens that a great proprietor is a great improver" and explains that

> The rapine and violence which the barbarians exercised against the ancient inhabitants interrupted the commerce between the towns and the country. The towns were deserted and the country was left uncultivated... During the continuance of those confusions, the chiefs and principal leaders of those nations acquired, or usurped to themselves, the greater part of the lands of those countries... All of them were engrossed, and the greater part by a few great proprietors.[29]

4.5.4 Greater Secularization of Power

In Byzantium, the Church never acquired primacy over the Emperor; exactly the opposite happened in the Holy Roman Empire, where the Pope's primacy over secular power was absolute. Furthermore, the Byzantine Church did not have the monopoly on education and could only establish religious schools. Already in the fifth century, the Emperor had established the "Pandidaktirion", which later under Justinian evolved into a famed law school. In the eighth century, the school introduced mathematics and astronomy, developing into a structure much reminiscing future universities.[30]

---

[27] An extensive network of post stations is founded as early as in the fourth century to facilitate communications in the Empire; see Papasotoriou, *Byzantine high strategy: 6th–11th centuries* (2001).

[28] According to Kordatos: "They were tied to their piece of land and were bought and sold with it... The serfs were a kind of crofters of the Byzantine era", *op.cit.* (Chap. A).

[29] Adam Smith, *The Wealth of Nations* (1986, Book 3, Chap. 2).

[30] For a history of the Pandidaktirion and the courses taught there, see Komninos, *Divine-built Walls* (1965).

References

Cipolla C (1994) Before the industrial revolution: European society and economy, 1000–1700. Norton, New York

Doyle EJ (1976) Two new fragments of the edict of Diocletian on maximum prices. Hesperia 45 (1):77–97. http://www.jstor.org/stable/147719

Gramsci A (1975) Prison notebooks. Columbia University Press, New York (Edited and translated by Joseph A. Buttigiegg, vol. I 1992 & vol. II 1996)

Harvey A (1997) Economic expansion in the Byzantine empire, 900–1200. MIET Publications, Athens

Komninos P (1965) Divine-built walls. Nea Estia 908, Athens (in Greek)

Kordatos Y (1928) The Thessalonika commune 1342–1449. Epikairotita Publications, Athens (in Greek)

Montesquieu (1968) Considérations sur les causes de la grandeur des Romains et de leur décadence. Garnier-Flammarion, Paris

Papasotiriou C (2001) Byzantine high strategy: 6th–11th centuries. Poiotita Publications, Athens (in Greek)

Rosen W (2008) Justinian's flea: the first great plague and the end of the Roman empire. Penguin, New York

Smith A (1986) The wealth of nations. Penguin, London

Spentzas S (1996) G. Plethon-Gemistus, philosopher of Mystra: his economic, social and fiscal views. Kardamitsas Publications, Athens (in Greek)

Stathakopoulos D (2004) Famine and pestilence in the late Roman and early Byzantine empire: a systematic survey of subsistence crises and epidemics. Ashgate, Aldershot

Tainter J (1988) The collapse of complex societies. Cambridge University Press, Cambridge

Economic Theories and Practices in Medieval Europe

5

Abstract

Feudal Europe witnessed not only the suppression of liberties but also the stagnation of economic thinking. For several centuries, lending and commerce were demonized as unholy activities, until an unprecedented disaster that took place during the four century in Europe shook the feudal structures and catalyzed a wave of political change and social mobility. Economic thinking was rapidly advanced so as to interpret the new realities of emerging trade and wealth accumulation. As taxation becomes less repressive and monetary corrosion more repulsive to the awakening masses, sovereigns seek alternative forms of borrowing from the nascent banking system. This brings about new financing opportunities for the rising national states and, when it becomes excessive, a new type of economic malaise: the debt crises.

5.1 Production and Economic Rules Under Feudalism

Feudalism emerged during Middle Ages in Europe, by destroying the multinational character of the Roman Empire and imposing a much more expanded and rigid version of community organization. Economies of scale were enhanced by the massive appropriation of lands by the new rulers, but population was subjected to severe working conditions and constraints of movement. Economic theories turned to strict recipes which subdued production and distribution to the needs of the sovereign, dismissing commerce and individual wealth. But taxation and serfdom continued to oppress the masses so as to ensure revenues and cheap labour for the landlords.

In the midst of chaos that followed the fall of Rome in 476, barbarian groups monopolized the lands of the empire and made the *fiefs*; these lands were not distributed or sold for cultivation but were inherited undivided by the eldest son (the primogenitor). Adam Smith underlines the fact that land descended undivided to the primogenitor was of tantamount importance for the security and preservation

N. Christodoulakis, *How Crises Shaped Economic Ideas and Policies*,
DOI 10.1007/978-3-319-16871-5_5

of the fief, a kind of economies of scale that allowed feudal landlords to establish their rule.[1] The indivisibility of large estates became the material basis of the feudal mode of production, in which the landowner also acted as the local ruler. The lands were cultivated by landless peasants, while an armed force was used to protect the fief both from raid by neighboring lords or domestic uprisings.

Most economic activity was based on limited or even barter exchange. Feudal lords (called '*seigneurs*') received part of the produce from their tenants and paid part of it to the King or other overlord; the remainder was either stored or sold. They had the sovereign right to mint their own coin (this is why the privilege of issuing currency is known to this day as *seigniorage*) which they used to pay their army and staff or import valuable commodities from abroad.

In Middle Age Europe, slavery is transformed into serfdom. Serfs were legally tied to the land they cultivated, but later they were allowed to move in order to find temporary employment in agriculture and husbandry.[2] Eventually, the system that prevailed was based on contracts between the serfs and the landlords, with the former renting the farmlands and paying in exchange part of their production to the lord; often, landlords would offer improved conditions and incentives so as to attract the most competent farmers. In late Middle Ages indigenous slavery disappears, with imported slaves from Africa gradually emerging as a vital part of the labour force.

The characteristics of feudalism were not universal; rather they depended on the region, where various forms of ownership and production can be found. The most powerful forms of feudalism emerged in France, though some communities managed to survive the barbarian raids and continued to claim ownership of their lands, often risking their lives. The conquerors' motto 'no land without a lord' ('*nulle terre sans seigneur*'), was never fully enforced thus some common lands coexisted with fiefs, preserving a few characteristics of the old communitarian way in the Middle Ages.[3] In Germany a variation of the feudal system based on the cooperative cultivation of the land prevailed. Farmers who lost their lands and rights of free movement became dependent on Teutonic warriors and lords, who provided safety and security. Contempt for commerce was widespread in Middle Ages, with the Church warning that most merchants will end up in hell. To mitigate this prospect, many of them left their riches to the Church which thus emerged as a great economic power.

The scale of production and the mobility of factors start to become less rigid by the eleventh century, since the first "guilds" were established as associations of artisans and professionals. This was the first step towards the emergence of producers as an independent social group, separate from landlords and merchants, a development which paved the way for the gradual transition from feudalism to a

[1] Adam Smith, *The Wealth of Nations* (1986, Book 3, Chap. II).

[2] Ladurie Le Roy, *Montaillu: Cathars and Catholics in a French village 1294–1324* (1981, pp. 116–117).

[3] See Anderson, *Passages from Antiquity to Feudalism* (1974, p. 148).

new system of economic production. Labour becomes more abundant at that time as many peasants abandon their villages to escape the marauding episodes of bandits and epidemics, and start looking for work in the cities. Guilds were strengthened as a result and the scale of production got enhanced further.

As Europe attempts to become a unified geopolitical entity, each state goes on a quest to discover new types of administration and representation. England has a Parliament which mitigates the absolutism of royal power, while in France the Estates General Assembly is established, a legislative body representing the different classes (estates) of the country including commoners (merchants and producers) as one of them; in any case the assembly is not as powerful as its English counterpart, lacking therefore the latter's stabilizing attributes.

Much of the economic thinking and teaching of the time focuses on justifying and expanding these trends. During the thirteenth and fourteenth centuries, economics were primarily a normative field, reflecting the subservient state of social organization in general. The thirst for knowledge is however insatiable, so the Church gradually relaxes prohibitions and by 1100, the first universities are established in Europe. They are initially founded as monastic seminars, gradually expanding to more structured knowledge-providing institutions.

The Scholastics, literate monks who taught at the seminaries, were essentially the first university professors. Their main sources of knowledge were Aristotle and the Roman law and, based on these premises, they dealt with various issues including commerce, interest rates, ownership, money and the perennial mystery surrounding the 'fair price' of goods. Since commerce constantly changed the economic conditions, as well as the scarcity and prices of goods, the Scholastics also studied the stability of money and its role in society. Critical thought is gradually awakening and the first principles of a more coherent economic theory finally emerged.

Thomas Aquinas (1225–1274)—a prominent philosopher of the Catholic Church with deep knowledge of Aristotle—tried to lift the moral condemnation of profit and argued in favor of private wealth. He argued that the price of selling goods could be allowed to vary in accordance with prevailing circumstances and still be fair. By making lengthy appeals to Aristotle, Aquinas gave his blessing to profit-making activities by reasoning that "gain. . .which is the end of trading, though it does not logically involve anything honorable or necessary, does not logically involve anything sinful or contrary to virtue;. . .and so trading will be rendered lawful".[4] However cautious such an argument may sound today it was ground-breaking for the ascending commercial activities and the legalization of the ensuing accumulation of wealth.

In his *Treatise*, Nicole Oresme (1320–1385) attempts to derive a theory on the function and use of money. In one of the first analyses on the role of gold and silver for the purpose of monetary stability, he argues that the speculative debasement of

[4] The excerpt is from the translation of Monroe, *Early Economic Thought: Selected Writings from Aristotle to Hume* (2006, p. 63).

currency is immoral and, therefore, 'it is scandalous and disgraceful for a prince to allow the money of his Realm to have no fixed value'. Later, however, Oresme conceded to the possibility of debasement to face unforeseen difficulties as, for example, "if the community has great need of a large sum of money for a war or for the ransom of its prince from captivity, or for some other emergency".[5] Oresme justified this deflection in the hope that the debased money would be transported into distant places, and obviously was indifferent to beggar thy neighbor effects.

5.2 A Century of Disaster and Change

In the mid fourteenth century, European countries suffered from a devastating plague pandemic, which resulted in radical social and demographic changes.[6] Human losses were enormous, as Europe lost half its inhabitants to the Black Death, and it took three centuries for population levels to recover. The plague also caused massive famine and an unprecedented economic and social crisis, which greatly tested institutions and values in all aspects of life.[7]

First, the economic consequences: the Black Death annihilated not only people, but also the livestock and a large part of cultivations. In France, the loss of agricultural production caused the quadrupling of food prices, while city workers rose up, demanding wage increases to cover the rising cost of subsistence. Despite efforts to suppress their demands, workers were often successful in getting significant wage increases from panicked employers.

As peasants massively fled to the cities to avoid the plague, landlords dramatically decreased rents and conceded a bigger portion of the produce as well as greater freedom of movement. In the cities, artisans who saw their assistants perish, opened the guilds to farmers who were looking for a job. Famine and pestilence became the catalysts for the rapid integration of farmers in the production process, the increase of wages and for breaking social norms that had persisted for centuries. On the dark side, many saw the crisis as just another opportunity to amass more wealth. Merchants and doctors demanded much higher prices for their products and services, and even priests asked exorbitant fees to perform funeral services. Many got married to sick individuals just to inherit their wealth and titles, while others forged wills and snatched up the fortunes of those who perished.

Many people lying on their deathbeds donated their wealth to the Church as a last plead to God, and bishops suddenly found themselves possessing vast lands,

---

[5] Excerpts taken from Monroe, *op. cit.*, p. 101.

[6] Much of the description in this section has heavily borrowed from Tuchman, *A Distant Mirror: The calamitous 14th Century* (1979).

[7] The simultaneity of famines and pestilences was frequent in antiquity, and this may explain why the terms for famine and pestilence are vocally identical in ancient Greek (*λιμός* and *λοιμός* respectively). The same phenomenon appeared later as well; an example is the 1258 AD famine in Italy, which was followed by a plague epidemic in 1259. See Nicholas, *The Evolution of the Medieval World* (1992, Part III, Chap. 12).

houses and gold. The power of the Church came under intensive scrutiny, as the clergy not only took advantage of the dying to usurp their properties, but had the hapless idea to interpret the plague as a divine punishment. To prove the theory of retribution, the Holy Inquisition blamed the Jews and unleashed a ruthless persecution against them, as was to happen several times afterwards. This dismal process of wealth redistribution deeply influenced medieval society and precipitated many of the developments that followed. Moreover, holding on to an irrational theory in times of a crisis only exacerbates the disaster. Church persecutions robbed society of experienced individuals who otherwise could have aided the sick, while obligatory participation in atonement litanies caused faster contagion. Several thinkers started to publicly air their discontent with absolutism and claimed more rights and freedom. In many respects, the economic and social annihilation of the fourteenth century became the catalyst for the transition to a new, freer and more open society.

The upheaval in economic relations caused by the Black Death was soon followed by a series of uprisings against the *ancien régime* composed of ruthless landlords, greedy lenders and the oppressive Church. Over-taxation was the most common root cause of the revolts. In Florence the revolt of *ciompi* (wool carders) broke out in 1378 against banks and harsh working conditions, and demanded representation in the local Government (*Signoria*). The insurrection was briefly successful for a period of 41 days, before being defeated. In England, the attempt of the King to enforce a new poll tax in 1381 (the third to be levied within 4 years) caused the Peasants' Revolt, with exhausted farmers and workers storming London and demanding lower taxes as well as political rights and protection from the arbitrary rule of landlords. Similar demonstrations took place in Belgium, while in 1382 the French crown's decision to impose new taxes on wine and salt caused a general revolt by workers, artisans and merchants in all major French cities. For many months, tax offices, churches and manors were looted. Finally revolts were violently and bloodily suppressed, but not uprooted. Their causes remained uncured and soon new waves of protest were erupted until feudalism was finally removed.

5.2.1 The Emergence of Trade and Banking

Plague and famine coincided with other major changes. Europe experienced sweeping climate changes in mid-fourteenth century, as the arctic ice melted, altering the entire ecosystem of Northern Europe. Polar temperatures in Scandinavian countries destroyed grain production, causing a great food crisis. These environmental changes intensified the search for new sectors of activity that would compensate the losses from the dwindling produce of the land.

As Central Europe was at the same time trying to compensate for losses in its own labour force caused by the Black Death, new production techniques were implemented, relying more on fixed investments as a way to substitute for scarce labour. This development gave birth to new opportunities for product specialization and market exchange. The frost on the farmlands in Northern Europe bolstered trade relations with southern countries and encouraged two-way migration streams.

The growth of trade was immediately followed by accumulation of wealth that called for further reproduction and growth, while at the same time more and more people throughout Europe demanded access to the new riches. Discontent was on the rise and sovereigns started to accept a more open system of representation and a more liberal process for acquiring and managing the wealth created by the new economic conditions. By the end of the calamitous fourteenth century, a new form of society had emerged, which had no tolerance for the old system but also lacked a clear vision of what should replace it.

It is interesting to note that this new economic blossoming was most profound in Italy, the country that was first hit by the Black Death, after Genoa had received the first merchant ships carrying the plague from the Black Sea. By then, however, the curse turned to eulogy: the Italian merchant fleet was the most convenient means for the growth of trade. In Florence and Sienna, trade fairs attract commodities and other forms of wealth from all around the known world, while the local production of wine and textiles is thriving. After the Church abolished the prohibition of interest, savings could be legally loaned, eliminating the need for usurers. This led to the establishment of the first banks in Italy, which issued loans not only to local producers, but even to foreign sovereigns. The opening and increasing availability of credit paved the way for the debt crises that would follow soon.

During late Middle Ages, the blossoming of cities, the popular revolts, the movement of populations and the reawakening of trade, contributed to overturning the economic and social stagnation of the previous period as well as developing new ideas for explaining the phenomena and guiding new developments. Economic thought became freer, developing in fields that were previously forbidden: commerce, savings, the price mechanism and of course the circulation of money.

5.3 Taxation and Sovereign Debt

The most characteristic developments during this period are the tax reforms in the economic front, accompanied by the expansion of rights and some nascent forms of political power sharing. These parallel developments were not a coincidence. Overtaxation was a major cause of popular revolts, while the potential of some kind of participation paved the way for negotiating the distribution of power between the rulers and the ruled.

5.3.1 Tax Crises and Representation

Despite the Church's sternness in the matter of interest-bearing loans and its long disgust concerning commerce, when it came to taxation its stance was unbending; no disobedience was allowed and full payment of the taxes owed to the ruler was demanded. Indeed, the medieval Church formed a close cooperation with sovereigns in tax collection, receiving part of state revenue and becoming very inventive in discovering new methods on how to increase revenues. The most famous source of Church income was the so-called "Absolution Certificates", which were sold to the faithful; the price was set according to the gravity of the sin committed and proportionally to the sinner's income. When the Vatican organized a mass sale of certificates to fund the construction of Saint Peter's Cathedral, it outsourced the collection of payments to the banks. The Vatican also gave part of the revenue to the Emperor, to fund his own projects and service his debts.[8]

The rising urban classes saw these practices as a pretext for the Church and the state to get their hands on their laboriously accumulated wealth; even worse, the revenue was being transferred to another country. Their protests paved the way for a new religious movement, "Protestantism", which from its inception made the stability of taxation and the efficient use of state revenue a crucial pursuit. Protestantism extols thrift, tax compliance and, in general, the fulfillment of a citizen's obligations without tricks or exceptions. German reformer Philipp Melanchthon invoked the phrase attributed to Jesus "render unto Caesar the things which are Caesar's, and unto God the things that are God's" to stress how important tax obedience is, thus providing the state with a mission compatible with the beliefs of the upcoming bourgeois class.

In its unceasing need for financial resources, the Church invented yet another kind of revenue, one that many sovereigns would later copy. In 1427, to address its dire need for cash, the Papal administration (*Curia*) sold the Holy See's Post Offices to private individuals. Several sovereigns—especially in France—imitated the auctioning of public offices and state functions in exchange for lump sum payments. The buyer took over the duties of the public institution and was allowed to make an annual profit to the tune of 8–12 %. In the case of taxation, tax collectors squeezed the peasants because whatever was left over, after meeting the obligation to the sovereign, went into their own pockets.[9]

The people were very hostile towards those private officials. In France an average of 38 revolts were recorded annually between 1662 and 1787, almost half of which were against tax collectors.[10] An additional source of grievance was the fact that many tax collectors were Jews hired by the sovereigns as economic advisors and managers, since—as non-Christians—they were not bound by doctrinal prohibitions concerning trade and lending. They were frequently targeted by taxpayers, but they also served as convenient scapegoats for the sovereign who, rather than letting his own greed for more revenues to be suspected, had no qualms about attributing the taxman's zeal to the fact that they had denounced Christianity.

---

[8] As Blissett notes in his historical novel 'Q' "bankers determined the manner of payment. Christians would donate for the construction of St. Peter's Basilica and would get in return a certificate signed by the Pope absolving them from their sins" (1999, Preface, Out of Europe, 1555).

[9] Blockmans, *A History of Power in Europe: Peoples, Markets. States* (1997, p. 227). Chapter 5 describes many characteristics of private authorities.

[10] Blockmans, *op. cit.*, p. 231.

5.3.2 Taxation in Late Middle Ages

In his treatise on the development of economic institutions, modern historian Niall Ferguson argues that the decision of the British Crown to cede fiscal control to Parliament during the fourteenth century was a landmark event in the interplay of politics with the economy.[11] It was prompted after King Eduard I had declared war against France and desperately needed cash to recruit mercenaries. The nobles, the rich and the landlords, all objected to the prospect of a new taxation and agreed to pay only if the King ceded control of revenue spending to their representatives in Parliament.[12]

They thought that only then would the King be prevented from misappropriating public funds, and state expenditure would become less wasteful. At the same time, the principle of universal taxation was enforced in England, whilst in other countries nobles had virtually no tax obligations. This expansion of the tax base fueled the demand for participation among the members of lower social classes and is the key factor behind many of the groundbreaking political reforms implemented at that time in England. Besides, it was in England where "everybody was equal before the law, there were no exceptions to royal taxation, and public offices were open to everyone regardless of birth".[13]

The Parliament, in order to perform its newly assigned role, hired specialized officials as civil employees—not private subcontractors—who paved the way for an organized tax collection mechanism. State revenue was no longer a monopoly of the sovereign, and this played an important role in the emergence of England as an advanced economic power. Indeed, many argue that a major advantage England had during her victorious wars with France was her ability to collect revenue for the war effort with minimal social discontent, since the English parliament's role in shaping the tax policy legitimized taxation. Even today, Britain is considered one of the most efficient countries in collecting and managing state revenue.

Things were quite different in other countries with regard to taxation and revenue collection. In France, the monarchy reserved the exclusive privilege of collecting and spending the income from taxes. Moreover, the crown privilege was sold to private individuals exacerbating public anger against tax collectors. Spain followed the English example and taxation was ceded by the King to more representative institutions. Already from the thirteenth century the Catalan Assembly had the privilege of collecting taxes and tariffs without much interference from the King. After the Bourbons came to power in 1714, the King had no jurisdiction even on the allocation of state revenue. The assemblies assigned the collection of taxes to Permanent Tax Committees, which consisted of judges, military officers and other state officials or private individuals. Committee members were paid from

[11] Ferguson, *The Cash Nexus: Money and Politics in Modern History 1700–2000* (2001, Chap. 3, p. 84).

[12] Later on, one argument used in favor of universal suffrage was the universality of taxation.

[13] See Lefebvre, *The French Revolution* (1962).

the proceeds they collected. These independent tax collecting mechanisms were eventually replaced by state bureaucracy in the nineteenth century, when absolutist kingdoms were transformed into constitutional monarchies and the assemblies evolved into parliaments. It was now the parliaments which had the power to approve the state budget and thus legitimize taxation.[14]

5.3.3 From Representation to Revolution

A similar *quid pro quo* happens elsewhere, with taxpayers achieving more representation and in exchange accepting new or higher taxes. Linking taxation with representation reaches its apogee in the United States, where in 1750 the citizens demanded that no taxation can be imposed without representation of the taxpayer in the legislative institutions.[15] This demand led to many conflicts with the British authority and was one of the major causes of the American Revolution a few years later.

In today's world, a characteristic case is Scandinavian countries with the highest average income tax burden in the world. It is no coincidence that these nations take top grades in matters relating to political representation, accountability and transparency. This ensures the efficient spending of tax revenue and legitimizes high taxes in the mind of taxpayers. On the contrary, south European nations are far behind their peers in average tax revenue. Often, this goes hand in hand with their lower performance in accountability and transparency of the central government.[16]

The negotiation between the government and taxpayers gradually lost steam and departed from its original goal of more political representation for a number of reasons. First of all, the establishment of independent tax authorities by governments deprived parliaments of the power to closely monitor the management of taxes or the collection and allocation of state revenues. Another reason was the expansion of political rights as an independent process. Since more and more groups were allowed to vote and be elected, citizens lost the traditional quid-pro-quo in their negotiations with authorities. Following this development, it is now the governments which take advantage of universal representation in order to legitimize taxation and its universal enforcement. Governments are eager to get the green

[14] Centuries later, Parliament's increased jurisdiction on fiscal matters often lead to widening public deficits. This led many to argue in favor of constitutional debt breaks, thus reducing the discretion of Parliaments.

[15] In 1885, Abraham Lincoln imposed new taxes and granted more equal representation, with the exception of Washington DC.

[16] Norway, Sweden and Denmark to the list of the 'Voice and Accountability' indicators of the World Bank with an index of 100, 99.5 and 99.1 respectively. Italy, Spain, Portugal and Greece trail below 80. The average tax revenue collected by central government in the first group is 28 % of GDP, while in the second 18.5 % of GDP. Data from World Bank, WDI (2012a) and WGI (2012b).

light from parliament every time they want to increase war spending or need additional revenue.

Nevertheless, the link between taxation and representation is not always straightforward. There are many cases where high tax rates are imposed by a heavy-handed state apparatus in nations where representation is minimal; this is the case, for example, with many authoritarian and dictatorial regimes. The opposite can also be observed, seeing a drastic reduction of tax rates when more representation is accomplished. The most characteristic cases are the nations of Eastern Europe which, following the collapse of communism in 1990 and the commencement of parliamentary democracy, adopted very low tax rates. Although their main argument for doing so was that lower taxes would attract foreign investors and bolster growth, one is also inclined to think that these decisions may have been influenced by the "de-legitimization" of state control that was harshly exercised in the years of communism.

5.3.4 Types of Taxation

One can make interesting historical comparisons of the types of taxation imposed, for example between taxes that are proportional to production and lump-sum taxes regardless of yield levels. A lump-sum has the advantage of motivating farmers to increase production and keep a bigger part for themselves. When drought takes its toll on farmlands, however, farmers turn over a bigger part of their produce and many sink into desperation. To avoid revolts, sovereigns applied some sort of "anti-cyclical policy", decreasing taxation during periods of drought, so that they maintain legitimacy when raising it again during prosperous times.

Payment in kind remained the most widespread form of taxation during Middle Ages as it had been in antiquity, since this system did not require elaborate estimates of the value of production or the circulation of money in the economy. But it also caused significant problems in logistics; besides the collection mechanism per se, another concern was the safety of transporting the produce from the more distant farmlands to the capital. The journey often took so long that the products arrived rotten in the capital, and this prompted the construction of roads to facilitate delivery; via Egnatia played a crucial role in the timely supply of Rome, while a road network with ample room for two carriages to avoid delays when they crossed paths was built in ancient China. After the collection, production was consumed by the ruling class and the army, and any surplus was sold at a price. In times of deprivation some part was allotted for free to the populace so as to prevent famine and avoid revolts.[17]

[17] One should not be mistaken to think that such phenomena existed only in ancient societies. In the modern era, famine revolts erupted in Paris in 1787, paving the way for the French Revolution of 1789. Exactly 200 years later in Romania, famished workers from the Braşov industrial area rose up against the communist regime; in 1989, they succeeded in toppling Ceausescu, who was sentenced to death.

The value of tax paid in kind was, however, liable to price fluctuations in cases of cyclical variations in production. To stabilize prices, the government of China applied a policy of intervention in the market, buying products when prices were falling and selling them when rising. Eventually, however, the Emperor decided that a more effective way to keep production and state revenue stable would be to replace payments made in kind with payments in cash. Since producers now had to sell part of their produce to find the money needed to pay taxes, the increased supply put downward pressure on prices and the tax burden increased in real terms.

When the tax was paid in kind, the quantity of goods was determined at will by the sovereign but nevertheless the result was a tax rate of roughly the same proportion across countries and historical periods. That was the *tithe*, or 10 % of production surrendered to the sovereign.[18] In antiquity, Aristotle gives an account of the tyrant Peisistratos visiting a village nearby Athens and hearing the farmers complain about their heavy labour and "*that's what Peisistratos ought to have his tenth of*".[19] Similar taxes were also imposed in the Byzantine Empire, India, France and elsewhere. During the Ottoman Empire, the tax was named "*dekati*" (deka = ten) for exactly the same reason. So what explains the popularity of this tax rate, especially if we take into consideration the fact that, historically, the average tax burden on income is around 30 %. A possible explanation may be that the produce of farmlands was divided into three equal parts, one payable to the landlord, another to cover the cost of the livestock, and the third was left to compensate the farmer. Therefore, a tax rate of around 30 % on the farmer's net income would be equal to the tenth of total produce.

5.3.5 Debt and Deficits

As raising taxes was gradually limited by the increasing representation, rulers became more prudent in managing revenues and expenditure during peace time. If the sovereign was in need of extra revenue to finance wars and other state functions, borrowing was the only alternative. However, this had its own pitfalls and soon the new practice was faced by a threat hitherto overlooked: the possibility of debt crises caused by states defaulting on their obligations. How did this new reality come about?

The accumulation of profits from trade gave rich merchants the leeway to lend money to sovereigns. In several cases, the loans were underwritten by issuing shares in the silver mines of America or claims on the cargos that arrived in Europe. When sovereigns were unable to pay their collateralized debt by employing such

[18] For an analysis of the tithe tax, see also Ricardo, *On the Principles of Political Economy and Taxation* (2004, Chap. XI).

[19] According to Aristotle, the tyrant appreciated this straightforwardness and granted him exemption from all taxes. That place was afterwards known as the "tax free farm". Aristotle, *Athenian Constitution*, (XVI, 6).

methods, they declared a default and negotiated more favorable terms with their creditors. Spain imposed such a payment moratorium nine times during 1557–1662, while France did the same three times. In many cases, state loans were underwritten with state-owned estates. When Gruyère, a county in Switzerland, defaulted in 1555, its lands were confiscated by the cantons of Bern and Friburg, the creditors, and remained annexed for two centuries.[20]

It was also during that period that the state became involved in business-supporting activities. The reason was that booming trade needed better transportation infrastructure and communication networks. The construction of new roads and bridges became a priority for many governments so as to advance activity in general and revenues in particular, but this required further borrowing. Rulers sought credit from abroad and soon countries were competing in providing loans to other nations. Holland, despite its small size, emerged as a great economic power of the time, becoming a major creditor by issuing loans at an annual interest rate of 3 % thanks to its ample monetary reserves from trade. Since in England the average interest rate was 6 % and in France 15 %, Holland quickly became the international financial center of the time.

Following the Revolution of 1688, the English Parliament decided to provide guarantees on state loans, making public debt a trustworthy asset and creating the first bond market. The establishment of the Bank of England a few years later (1694) meant that the issuing of currency would now be supported by reserves of precious metals, while the money market was also rapidly developing toward its modern form.

5.4 From Autarky to Trade

Circa 1500, a number of major changes sweep the economy. Peasants abandoned the countryside and looked for work and more freedom in the cities. Guilds expanded and trade was growing. The urban population becomes wealthier and demands political power undermining the status quo of the time, that is the alliance between the Church, the great landowners and the army. New lands and a new continent are discovered. The circumnavigation of Africa opened up new merchant routes to India and China, while European nations who conquer distant lands, assume access to their products and natural wealth and emerge as new world powers. The rapid growth of trade is followed by an impressive accumulation of wealth, mainly from commerce.

Wealth was not compatible, however, with the old norms and beliefs. A new paradigm was in search to provide moral and theoretical justification to private property and accumulation. New institutions were needed to protect it as this new wealth and the new commodities fueled envy and violence and were accompanied

[20] As described by Oechsli, *History of Switzerland 1499–1914* (1922, p. 128). Following the principle *cujus regio ejus religio*, Bern also imposed the Reformation upon its spoil; (*ibid*).

by rising crime rates. For many people, looting became a more attractive choice than working and producing, and the first post-feudal societies sunk into chaos. Events seemed to vindicate what Roman poet Lucretius, reflecting on the dissolution and restructuring of societies, had written many centuries earlier wondering that:

> Thus things down to the vilest lees of brawling mobs succumbed,
> whilst each man sought unto himself dominion and supremacy.
> So next some wiser heads instructed men to found
> The magisterial office, and did frame
> Codes that they might consent to follow laws.[21]

It is under these circumstances that the first theories concerning the establishment and empowerment of state institutions and the need to impose rules on economic behavior start to emerge. Economic theories were also tasked with defining concepts such as self-interest and addressing the problems of utility maximization and selection of means to achieve it.

5.4.1 Theories of Money and Trade

Goods and commodities are not the only ones to travel overseas; trade allows the dissemination of ideas, habits and behaviors, often subverting and occasionally overturning the prevailing ones. For example, the price of products can now be determined by the law of supply and demand instead of following the guidelines of the Church regarding "fair price". Even money, the accumulation of which was condoned by the Church only in the case of the sovereign, started to circulate, being exchanged and deposited. The trend of '*bullionism*' (bullion = bar of precious metal) was propagated among the rising social classes of the time. The hoarding of precious metals allowed the rich to avoid both taxation and the wrath of the masses, had they turned all this wealth into luxurious consumption. In this way, they also protected their fortunes from currency debasement through clipping, a practice increasingly popular among sovereigns.

The discovery of America showered Spain with riches and outlets were desperately sought for the new wealth to be invested and multiply. Several economic thinkers in Spain, becoming known as the Salamanca School, advocated that profit-making from transporting and reselling goods is morally acceptable, calling for the Christian bias against merchants to cease. Since money was such a commodity, it was entitled to demand an interest rate when loaned; this put an end to the medieval prohibitions of lending at interest. During the same period, new and bigger banks are established to manage the money made by commercial activity, issue loans to merchants and keep their deposits safe. As the first banks had been founded in the fourteenth century close to the trade fairs of Florence and Sienna, new banks appear in places where commerce was thriving and money accumulated. Good cases in

[21] Lucretius, *De rerum natura* (1981, Book V).

point are the Transaction Bank (*Banco de Giro*) established in 1619 in Venice, and the *Wisselbank* established in 1609 in Amsterdam.

Examining the changes brought about by the growth of trade in individual and collective behavior, Manent writes that, until then, the defining characteristics of human societies were commanding and obeying, but now

> ... [a] new order is born... no longer based on commands and orders, but on what has been called, with increasing enthusiasm since the 17th century, self interest... [I]n this new order, motivation for action is inherent, with the individual applying reason to determine what is good for himself and his freedom and seeking the most expedient means to achieve his goals.[22]

5.4.2 Class Oriented Theories

New economic thinking now openly acknowledges that trade bolsters wealth, growth and power. In England, Thomas Mun (1571–1641) argues that trade surpluses are a prerequisite to a nation's prosperity and power. Despite their internal contradictions, these views prevailed in many nations for almost three centuries, constituting the economic doctrine of "mercantilism" (from the Latin god Mercury, protector of trade).

Commerce, however, did not just increase the revenue of the sovereign from tariffs and taxation. It also fueled the demands merchants had from their government, leading to a new paradigm with regard to the role of the state in the economy. For example, the demand that the state should protect trade routes led to the gigantic expansion of national fleets. Pressure for controlling new lands, rich in raw materials and new markets for European products, led to conquests and colonization; the pressure for advancing domestic production in order to sell commodities in foreign markets became a motive for massive investments and mass employment.

These developments also raised the issue concerning the structure of the domestic economy and open rivalries festered among various groups: theorists were asked by interested parties to provide rules that could help them and arguments they could use against their adversaries. The estate classes found themselves sidelined by the new economy, and their reaction in France gives birth to the theory of Physiocracy. In England, when the rising class of producers thought that merchant profits had become excessive, their contestation led to the theories of savings and capital accumulation developed in Scotland in the eighteenth century. Both theories shared their deep contempt for mercantilism, conveniently forgetting of course that the gigantic growth of agricultural and industrial production they had enjoyed was brought about by the preceding growth of trade.

---

[22] Manent, *Cours familier de philosophie politique* (2004).

5.4.3 Church Reform

This new balance of economic power and the demand for more political rights seriously subverted the authority of the Church, leading to Luther's protest in Germany in 1517. The movement of Protestant Reformation formulated a new doctrine of individual morality, complemented by a new set of rules concerning the economy and, more specifically taxation and lending. Luther castigated "those who should be governing the country and the people, as deviating from their mission and—in contrast—they rob people…imposing tax upon tax and tariff upon tariff".[23]

The individual is no longer thought to be subservient to divine power or—by extension—to God's representative on earth, the Church. Individuals are now responsible for their own actions and for the consequences of these actions, which determine whether they will be saved or dammed. New ethics are also established regarding labour; work is no longer considered an imposition by the sovereign on his subjects, but a personal duty and a crucial factor for individual and social progress. Such an approach echoed the teachings of Aristotle, who had argued that individual as well as collective political and economic behavior should be appraised on ethical grounds and comply with moral standards.

5.4.4 Social Positivism

The new approach was the in the opposite of theories viewing society as the playing field of conflicting interests and searching how a balance could be achieved without moral coercion. The most important representative of such political "positivism" is Machiavelli (1469–1527), who examined the role of the state and the coexistence of constituent groups based on their differences and argued how they can still serve the "common good". Machiavelli's analysis entails three critical elements, such as (a) an empirical analysis of reality without moralizing, (b) explicitly stating the goals pursued by each group, and (c) identifying the means for achieving them by taking into consideration the constraints faced by each group. This type of analysis, as well as the distinction between individual and general good is common in many later theories of political organization and economic behavior.

Despite his path-breaking approach, many would be eager to paint Machiavelli as a philosophical *prince of darkness* by interpreting his work as legitimizing any actions that serve a specific goal—aptly expressed in the aphorism "*the end justifies the means*".[24] However, this interpretation is unfair to Machiavelli; nowhere in his work does he provide any moral legitimacy to the means used, and probably the view was due to confusing positive analysis with normative advise. He just revealed and described the actual workings of power, not touching on issues of ethics and

[23] From Luther, *Von Welltlicher Uberkeytt*.

[24] In fact, this was again a Jesuit principle; cf. Section 2.1, footnote 3.

morality. The real motive behind the criticism was rather that authorities were enraged that he had ripped apart the veil of moralization which masked their own political action and conceal the devastating consequences on the citizens.

Machiavelli's anti-authoritarian and democratic principles are manifest in his views about the structure of the modern state, where he declares that the multitude is wiser, more constant and more just than a prince.[25] Consequently, he argues that all power should be based on citizens, who, contrary to a prince, will never elect to public office a man with a bad reputation or record of corruption. Machiavelli is also one of the first theorists to distinguish between individual motivation and collective behavior, and noted that collective action is not always the sum of individual actions. Commenting on an incident from Roman history, when the Senate had ordered some Roman fugitives to return to Rome under serious penalties, Machiavelli writes that "all together they are strong, but when each one then begins to think of his own peril, they become vile and weak".[26] This view is held by many modern economic theories dealing with the issue of diverging individual motives in big groups.

5.4.5 The Advent of Reason

Challenges to the old regime cut through every aspect of society. In science, Copernicus and Galileo established the heliocentric theory, according to which the earth is not the center of the universe but just another planet orbiting the sun. The consequences far exceeded those of a scientific discovery. When the dogma of Earth's immobility, which was the basis of the theological interpretation of the world, crumbled, the same fate awaited the supremacy of the Church and the authority of kings to rule by divine right. Nothing seemed stable or certain anymore and tectonic shifts would soon follow.

Besides celestial phenomena, thinkers were also occupied with earthly matters, trying to develop new theories and interpretations of the world. In England, a Franciscan theologian, William of Ockham (1288–1347), had argued in favor of the reason and not theological doctrines. In his desire to rid sterile and pretentious philosophizing, he coined a principle which became known as "Ockham's razor"; according to which if two theories explain a phenomenon equally well the one with the fewer assumptions should be selected. In economics, the principle proved to be very useful in discarding complicated assertions that hope to impress the laymen exactly by not being fully understood.

The introduction of empirical analysis was met with great enthusiasm in economic research. Actual observations, classifications and comparisons gradually displaced previous abstract teachings and arbitrary doctrines held to rationalize economic behavior. The same liberating effect was felt in politics, this time as a

---

[25] Machiavelli, *The Discourses* (1988, I. 58, p. 255).

[26] Machiavelli quotes the Roman historian Livy, *op. cit.*, I. 55, p. 250.

result of the collapse of the awe-inspiring geocentric model. In the Middle Ages, self-interest was considered tantamount to sin (one exception, of course, being that of the sovereign), and individual behaviors should conform to the norms and admonitions that ensured the balance and stability of the feudal status quo. This changed; being released from the fear and suppression of absolutism, people now pursued their individual goals more freely.

The process, however, was not sweeping enough and it took three more centuries for scientific reasoning to prevail. In France, Descartes (1596–1650) systematized mathematics and introduced the theory of logic. His work praised the value of scientific inquiry and permanent doubt in the process of seeking the truth. By his method, Descartes aimed at two things: First to set up a system replacing the scholastic tradition which he thought was no more than a stock of platitudes tailor made to appear in conformity with authorities.[27] Second, he set one's freedom as a critical presupposition in search of the truth, for otherwise the combination of ignorance and repression will harm the interests of society. Descartes considered that freedom cannot be traded off with material advancement and "put among the excesses all the promises by which some part of one's freedom is taken away".

References

Anderson P (1974) Passages from antiquity to feudalism. Verso, London
Blissett L (1999) Q (Ecclesiasticus). Einaudi Editore, Torino
Blockmans W (1997) A history of power in Europe: peoples, markets, states. Fonds Mercator Paribas, Antwerp
Descartes R (1968) Discourse on method and the meditations. Penguin, London
Ferguson N (2001) The cash nexus: money and politics in modern history 1700–2000. Penguin, London
Le Roy Ladurie E (1981) Montaillou: Cathars and Catholics in a French village 1294–1324. Penguin, London
Lefebvre G (1962) French revolution. Columbia University Press, New York
Lucretius (1981) On the nature of the universe (De rerum natura). Penguin, London (Metrical translation by W. E. Leonard, Project Gutenberg)
Machiavelli N (1988) The discourses. Penguin, London
Manent P (2004) Cours familier de philosophie politique. Gallimard, Paris
Monroe AE (2006) Early economic thought: selected writings from Aristotle to Hume. Dover, Mineola
Nicholas D (1992) The evolution of the medieval world: society, government and thought in Europe 312–1500. Longman, London
Oechsli W (1922) History of Switzerland 1499–1914. Cambridge University Press, Cambridge
Ricardo D (2004) Principles of political economy and taxation. Dover, New York
Smith A (1986) The wealth of nations. Penguin, London
Tuchman B (1979) A distant mirror: the calamitous 14th century. Penguin, London

[27] Descartes, *Discourse on Method and the Meditations* (1968, p. 46). He was deeply critical of those following Aristotle though not possessing as much knowledge of nature as he had, and described them "...like the ivy which does not seek to climb higher than the trees which support it", (p. 83).

World Bank (2012a) World development indicators. http://data.worldbank.org/products/wdi
World Bank (2012b) Governance indicators. http://data.worldbank.org/data-catalog/worldwide-governance-indicators

Economics Before the Industrial Revolution 6

Abstract

At the dawn of the modern era, the combination of political liberties and rational thinking was quickly undermining the dominance of religious doctrines and superstitions. Social order could not be any more imposed by ruthless repression, and political philosophers set out to propose a new balance between individuals and the state with various degrees of state empowerment and individual choice.

As deep conflicts of interest emerge among the protagonists of the new economic order, rival economic theories are formed to defend them: mercantilism supports the expansion of trade as the only means for a nation to accumulate wealth, while Physiocracy argues in favor of land owners and producers. The ceaseless quest for overseas investment opportunities and the increasing availability of credit lead to speculative bubbles and the outbreak of the first financial crises. Governments are puzzled on whether individual and collective interests can ever be compatible and new challenges open up for economists. On another ongoing crisis, however, authorities and thinkers were less eager to take up the challenge and, in sharp contrast to the progression of liberties in Europe and North America, the suppression of human labour in the form of slavery was often justified.

6.1 The Political Economy in England

The uncertainty surrounding the outcome of the awakening and mobilization of previously subordinate and subjugated classes terrified the old regime and raised significant questions for philosophers and thinkers. Just as scientific inquiry had proved that heavenly bodies follow regular paths without colliding and chemical particles irregular ones without disappearing, the philosophers of the time attempted to analyze the new social dynamics. Working from the individual to the general, they tried to identify how conflicting self-interested behaviors are balanced. In England, the two major philosophers of the time were Thomas Hobbes

N. Christodoulakis, *How Crises Shaped Economic Ideas and Policies*,
DOI 10.1007/978-3-319-16871-5_6

and John Locke. Their political philosophy would become the basis and source of inspiration for later economic theories regarding the establishment of the state, the role of property and labour, and the structure of society in general.

6.1.1 Hobbes's Theory of State

In mid seventeenth century, Thomas Hobbes (1588–1678) argued that a powerful state is founded as a guarantor of individual activity. He insisted that the state should make the rules and supervise their enforcement, so that individuals can coexist peacefully, avoiding mutual destruction in order to gain power. His theory is the exact opposite of the "*median*" theory of Aristotle, according to which excess is considered as harmful as deficiency. Hobbes, in contrast, argues that mankind has an inherent inclination in seeking power that ceases only in death. For Hobbes, the building blocks of society are the diverse passions of its members, uncontrollable sentiments and self-interested behavior, and to survive means to always demand more.[1] There is no point in trying to embellish or dismiss reality by moralizing doctrines of vain philosophy, as he scornfully characterized the Aristotelian teaching.

The heart of Hobbesian theory is the commonwealth, which can accomplish its moral goal—and in this he agrees with Machiavelli—without being necessary that all members of society share the same values. Quite the contrary: Hobbes ferociously dismissed both the metaphysical doctrine that human actions are dictated and ultimately judged by God, as well as the approach of scholastic philosophers, who attempted to refine this imperative with the concept of some absolute and ubiquitous morality. To resolve the problem of unconstrained behavior, Hobbes believed that harnessing it for the interest of society is justified, but this should be based on the observations of experience and scientific knowledge, not vague moralism.

This begs the question: which institution is to be tasked with observing and serving the commonwealth? Hobbes's answer is that this can be accomplished only with a powerful state and uses the biblical sea-beast *Leviathan* as a metaphor for its size and awe-inspiring character. The state is responsible for organizing social behavior, ridding society of the ills of domestic strife and defending itself against external threats so as to safeguard the domestic wellbeing. In dealing with the potential divergence between collective goals and private interests, Hobbes resolves the issue by accepting individual freedom, but only as the result of rational thought in the context of a social environment rather than the product of reckless behavior. According to Hobbes, being free means that the individual uses his

[1] Hobbes, *Leviathan* (1982, First part, XI). Due to the endless quest for domination, according to Hobbes, the doctrine of right and wrong is perpetually disputed, both by the pen and the sword. He jokingly remarks that if Euclidean theorems could be challenged by those in power, then all books of geometry would be burned and suppressed.

abilities and mind to do what must be done. Freedom is exercised even when the individual makes choices he does not like; for otherwise the consequences of inaction could be even worse (for example one should pay taxes, lest he is punished). No wonder that Hobbes is considered the father of modern political theory of the state.

Hobbes hugely influenced the course of normative economic thought, providing the philosophical underpinnings for collective well-being, formulated later by Bentham in the nineteenth century. However, Hobbesian philosophy also had many critics. In France, Voltaire reinstated the limitation on individual and collective action, arguing that one's freedom ends where another's starts. Later, Pareto would famously argue that there is only one criterion of legitimacy in any change in the relative position of parties: that no one is made worse off and at least one is made better off. The syllogism rendered Hobbesian authoritarianism into a potential threat against one's prosperity, and the challenge was taken up a few years later by another English philosopher.

6.1.2 Locke's Political and Economic Theories

John Locke (1632–1704), was a leading political thinker who established the philosophical system of positivism. Living in a time of greater freedom and upheaval compared to the era of Hobbes, Locke considered the individual not as a source of uncontrollable behavior, but as the agent of justice. Nevertheless, he recognizes that it is difficult to define the "common good" as a sum of individual interests. Locke comes to accept that the state should have the authority to make decisions and act on behalf of society, though this prerogative should never be unconditional or limitless as proposed by Hobbes. According to Locke, the state's power to act using discretion for the benefit of the community is undoubtedly prerogative,

> . . .for the people are very seldom or never scrupulous or nice in the point or questioning of prerogative whilst it is in any tolerable degree employed for the use it was meant—that is, the good of the people, and not manifestly against it.[2]

According to Locke, such an evaluation is not the task of the state alone but of a "social contract", which guarantees the natural rights of the individual. The state need not be Hobbes's *Leviathan* but the agent of citizens who must remain watchful and monitor the credibility of government. If they judge that the government no longer fulfills its mission, they have the right to overturn it.

Locke also studied several economic issues and examined their role in serving one's interests. According to Locke freedom, along with life and material possessions, constitute a man's wealth, which is inviolable. From a theoretical perspective, his most important contribution is the theory of property, on which

[2] Taken from Locke, *Second Treatise on Civil Government* (1690, *On property*, 34).

he subsequently formulates a labour theory of value. These contributions are the foundations of political economy and of the modern philosophy of rights in rem.[3] According to Locke, God gave the world to men *in common* for their benefit and use. This did not mean, however, that God intended the resources to remain unutilized just in order to be common. They must be used by the industrious and rational, and not for the fancy of the quarrelsome and contentious. At this point Locke follows Aristotle who had argued that the allocation of goods and wealth should be based on the citizens' worth rather than ancestry and inheritance.

Many accused Locke of raising private property to the status of a natural right, but what he meant by wealth was closer to what we would term today as human and physical capital. Indeed, Locke stressed that Nature's measure of property is set by the extent of men's labour and the conveniences of life; no single individual could appropriate them all, nor could his enjoyment consume more than a small amount, lest he encroaches upon the rights of another. Nor should a man acquire to himself the property of his neighbors, who would still have room for as good and as large a possession as before it was appropriated. This natural measure implies that every man's possessions are confined to a moderate proportion and such as he might appropriate without injury to anyone else. Again, a reflection of Aristotle's "*median*" criterion.

Emanating from the belief that the accumulation of wealth based on individual labour is the cornerstone of society, Locke comes to the conclusion that the value of a commodity is related to the labour needed to produce it. In early nineteenth century, the labor theory of value was used by English radicals to challenge inequalities in production and distribution of goods. Later, it would become the main analytical tool used by Marx in his theory of surplus value and class exploitation.[4]

Since labour has already been incorporated in goods, a commodity's value should remain stable, for otherwise the effort to produce it would be wasted. This conclusion leads Locke to seek an "anchor" in the economic system, in order to avoid a drop in value that would cause a *post facto* devaluation of labour. Hence, he formulates the quantitative theory of money, the velocity of circulation and the rules for issuing currency, all of which pave the way for the eventual emergence of modern monetary policy. Locke did not subscribe to either of the two dominant theoretical doctrines of his time, namely Physiocracy and Mercantilism, and although in favor of agricultural production he was not against commerce and trade.

---

[3] For an analysis of Locke's contribution, see Kitromilides, *Political thinkers of the Modern Era: Biographies and Interpretations* (1998, p. 84).

[4] See, *The Locke Reader* (1977, p. 45).

6.2 Physiocracy in France

The new economic realities brought about deep divisions in French society. The King and nobility continued to tax the farmers excessively and impose high tariffs on imported goods, thus causing severe shortages and famines. Many revolts against the crown took place in the countryside, while the rising merchant class was competing against the landlords. Economic thinking quickly provided the theoretical underpinning to these divisions. It was Minister Jean-Baptiste Colbert under Louis XIV who emerged as an ardent supporter of mercantilism, insisting that growth of trade is the key to accumulation of wealth. Consequently, reaction against the primacy of commerce arose among those who were losing power and influence: landlords and farmers.

6.2.1 Physiocracy

The theory of Physiocracy favors the agricultural sector and the primacy of land; industry and trade are considered as sterile activities that only recycle nature's production. Richard Cantillion (1680–1734), an Irish-French merchant and economist, was one of the first systematic thinkers on trade. In a study on how market prices are reached, he described a mechanism of continuous *altercation* similar to that of *tattonement* developed two centuries later by Walras. He also pointed to the inflationary impact of monetary expansion by stating that if [more] "...money is taken from the mines, the prices of all things will be increased by this abundance of money". Cantillion refrained from granting commerce any major role in the economy, as he thought that land is the only real source of wealth and hence landowners should be the ruling class in a society, since *the Proprietors of lands are alone naturally independent in a State; all other orders are dependent.*[5]

In the same spirit, Francois Quesnay argued that agricultural production is the only determinant of economic growth; to illustrate the point he published in 1758 his *Tableau économique*. There he describes the flow of production and/or transactions between the three classes of the economy (agricultural laborers, landowners and merchants/artisans) and demonstrates how agricultural labor and capital (livestock, raw materials and installations) confer value to manufactured goods. The Tableau is considered as the first attempt to quantify the flow of money and goods in an economy and has since been used in many cases of economic analysis.[6]

[5] Taken from Monroe, *Early Economic Thought: Selected Writings from Aristotle to Hume* (2006, p. 261).

[6] Marx's friend and coauthor, Friedrich Engels, was so impressed by the Tableau that he attempted to apply Quesnay's pattern in studying the structure of society into three classes; namely those of producing, of appropriating the surplus, and the industrial-professional class that is sterile because it simply adds only as much value as it consumes in means of subsistence; see Engels, *Anti-Dühring* (1947, Part II, Political Economy).

On a more practical level, the policy measures suggested by physiocrats had the single goal to support farmers, and included:

(a) Lower taxes on farmers, who should also have the right to choose their crops without state interference.
(b) Abolition of price-caps on agricultural products. These were used from time to time by the government to make commodities cheaper for the urban population.
(c) The abolition of all incentives for expanding the industrial sector, since these lured the workforce to the cities thereby undermining agricultural production.

Physiocracy influenced important strands of the French Enlightenment, highlighting its political and social dimension. The most militant was Jean Jacques Rousseau who identified the natural liberty of men (as per Locke) with the supremacy of nature as defined by the physiocrats. Rousseau also proposed a social contract, based on nature's dictates, far from social and political coercion. While in Locke the social contract is the basis for one to attain and exercise the right of liberty, in Rousseau is the framework to contain the individual from surpassing the limits set by Nature.

A central theme in Rousseau's philosophy concerns the insurmountable contradiction between mankind's natural state of innocence and morality, and the alienation imposed by social structures. According to Rousseau, the culprit for this regrettable development is technology which endlessly gives rise to new needs and complicates the social structure. Men are thus obliged to depart further from the "Great Conciliator", that is Nature, which for Rousseau represents the source of the unique and supreme good.

His religious physiocracy leads Rousseau to condemn any attempt of man to use technology to control or exploit nature. Shocked by the rapid technological changes around him, he points out the never-ending upheaval they cause to the human soul: "[e]verything here on earth is in a constant flux which allows nothing to assume any constant form" he writes to describe the deep uncertainty of his age.[7] Nothing remains stable". Rousseau's work represents perhaps the most explicit expression of "*technophobia*", by blaming technological progress for the economic and social ills plaguing mankind.

The greatest opponent of physiocrats was Henri Saint-Simone (1760–1825), a fervent supporter of scientific discovery and industrialization as the forces that could liberate society from the bonds of absolutism and enslavement to landlords.[8]

A symbolic landmark in the conflicts between physiocrats and Saint-Simonians was the destructive earthquake that hit Lisbon in 1755 on a Sunday morning, when

[7] Rousseau, *Reveries of the Solitary Walker*, Ninth walk (1981).

[8] Saint-Simon also defended the rights of workers and is considered as the father of 'utopian socialism'.

most victims were attending mass.[9] All agreed that the event was a decisive point in European history, but interpretations varied: the Church saw it as an act of divine retribution, intended by God to chastise the flock's depravity and lewdness. Rousseau saw it as nature's revenge, for man's greed had led to the erection of modern Babel Towers. The Church's adversaries argued that the earthquake was proof of lack of divine providence, since God had failed to save even His most faithful. Finally, supporters of scientific progress saw the earthquake as a challenge to improve construction techniques.

6.2.2 Dissemination of Physiocratic Theories

From a philosophical point of view, physiocrats believed that the state neither can nor should attempt to regulate or guarantee the proper functioning of the economy, and their motto was "*laissez-faire la nature*". This looks similar to many other theories arguing about the supremacy of agricultural production and the need for society to be placed under the limitations and primacy of nature.

In this vein, several communal movements that emerged a century later as a reaction to industrialization and capitalism were inspired by the preaching of physiocrats. In England, Malthus formulated his relentless theory about the dangers of overpopulation inspired by the doctrines and aphorisms of Physiocracy. It also motivated economic approaches prevailing in Germany at the time, and advocated agricultural production based on communal and cooperative structures. Similarly, the movement of German romanticism claimed that people can live in harmony only by restoring balance with Nature. Today one can easily identify the key tenets of physiocrats with certain radical environmental approaches, such as the techno-phobic aversion for industrial activity and technology advancement. A further demonstration of how certain economic ideas reappear in history, one can compare the measures suggested by physiocrats to support farmers with the Common Agricultural Policy implemented in the European Union. The CAP not only imposes tariffs on agricultural imports to the EU but also directly subsidizes the agricultural sector, whereas other sectors of the economy are deregulated.

Thus, it comes as no surprise that physiocrats were fiercely attacked by the supporters of industrialization: prominently among them was Adam Smith who wrote a philippic on the impasse of their policy prescriptions. Other critics included German philosophers and writers, notably Goethe who fervently urged the use of science for the liberation of man from "nature's bondage". At a philosophical level, Immanuel Kant (1724–1804) dismissed Rousseau's physiocratic thesis on the presumed innocence of a man's natural state and, in agreement with Hobbes, argued that not only does society not destroy man but, on the contrary, acts as a civilizing force. What is needed is the prevalence of public or general will that provides

[9] For a detailed account of the Lisbon earthquake and reactions at the time, see Shrady, *The Last Day: Wrath, Ruin, and Reason in the Great Lisbon Earthquake of 1755* (2009).

security to all; hence each individual should limit his freedom so as not to trespass on others.[10]

To this effect, Kant stressed the importance of individual moral responsibility which is necessary for balancing the relationship of individuals with society; thus men can dismiss absolutism without descending into a state of dissolution and mutual destruction. Men's actions should comply with the 'categorical imperative' endorsed by their own free will as a set of individual rules of morality. Contrary to Machiavelli, Kant argued that the only measure for all actions is the morality of the motive and not an outcome-based utility; indeed, the categorical imperative is blind to the outcome of actions.

6.3 Machines and Measurement

During the seventeenth century, England had built a powerful merchant fleet, thanks to which it eventually ruled the oceans and dominated overseas trade. In turn, this revealed the great potential of the manufacturing sector far beyond the capacity of artisans and guilds, thus new methods for increasing productivity and the scale of production were fervently sought.

6.3.1 Engineering as an Economic Force

At the same time great advances in mathematics and natural sciences were taking place, leading to technical discoveries that paved the way to the Industrial Revolution. The most prominent figure of the period was Isaac Newton. His law of gravity and the corpuscular theory of light had profound effects, not only in advancing scientific knowledge but also by shattering long held superstitions on the movement of heavenly bodies and the emission of light. Besides, Newton took the initiative for the establishment of the Royal Academy of Science, an institution designed to promote new discoveries in all scientific fields. This had far-reaching consequences as it liberated scientific inquiry from the censorship and control of the Church, and paved the way for further experimentation and inventions. Royal endorsement caused a public stir and discoveries became a pursuit of the many rather than the secret of the few.

In more applied fields, scientists studied kinetic theory and the theory of heat in an attempt to invent machinery that could augment production levels. One of the most important inventions of the time was the steam engine by James Watt, perfecting a previous piece of machinery built by Cartwright for the textile industry. By dramatically reducing costs, the steam engine revolutionized the whole industry and allowed England to overtake Holland in textile production. The eventual

[10] Kant, *Political Writings* (1977, p. 27).

introduction of the steam engine in all sectors of production, transportation and shipbuilding, created the Industrial Revolution.

6.3.2 Measurement as an Economic Tool

Following groundbreaking changes in production, technology, social organization, political philosophy and the sciences, economic theory was founded in the eighteenth century as a coherent system of thinking. The interaction with other sectors inspired a boom in new economic ideas with two novel characteristics: measurement and empirics.

The emergence of empirical observation as a tool of knowledge ended the reign of scholasticism according to which theories should be developed only from accepted premises and axioms. Until then, thinkers first interpreted reality in philosophical terms, and then attempted to put it in some quantitative form, if at all. In the new environment, economic theories started to develop on a quantitative basis by attempting to explain real-life phenomena. The English economists of the time followed the opposite direction: measure first, explain afterwards.

The first attempts included systematically measuring economic activity, observing changes in economic fundamentals and comparing the effects of various policies. Economic records did exist before but were kept mainly for tax reasons, while censuses were used primarily for military conscription. William Petty was the first to make a rigorous assessment of England's wealth in his *Political Arithmetic*, published in 1670. Adopting Lockean concepts of value, he considered the worth of a nation's wealth as stemming from cultivated lands, physical capital used in production, and human capital representing labor. Assuming that a layman's productivity is measured by his remuneration and taking the number of years an individual works, Petty estimated that the value per laborer is 20 times his annual income. If we assume that future income is expected to remain at current levels, Petty's calculation implies an annual 5 % discounting rate—close to the level used today in similar calculations.[11] A few years later, Graunt and King would study the laws of population growth and develop the first system of National Accounts.

6.4 Early Economic Crises

6.4.1 The Dutch "Tulipmania" (1636–1637)

Despite its small size as compared to its neighbors, Holland had emerged as a great seafaring power, taking advantage of both the demise of erstwhile city states (e.g. Venice) as well as the geopolitical vacuum left by other powers after the

[11] If the current and expected income is (Y) annually and the discounting rate ($r = 0.05$), the present value of lifetime labor is obtained as $V = Y + Y/(1+r) + Y/(1+r)^2 + \ldots = Y/r = 20Y$.

Thirty Years War. By accumulating wealth and building a strong fleet, Holland won naval supremacy by achieving victories over Spain as well as England, the latter being further weakened from a civil war.

Circa 1600 Holland, thanks to the huge inflow of gold from its colonies, had become a financial heaven, boasting immense cash reserves and easy access to bank credit. The government had allowed the operation of private mints in which cargos of bullion were turned into money without much scrutiny regarding their origin. This abundance of wealth caused the so-called *tulipmania*, the first commodity market "bubble" on record. In just 6 months—from November 1636 to May 1637—tulip bulb prices skyrocketed and then suddenly collapsed. Though the crisis did not seriously affect overall economic activity, the causes of the bubble have been a source of fascination and tulipmania became a timeless symbol of reckless markets—and market bubbles would surely burst many times since then.

Three theories have been suggested to interpret tulipmania and, by extension, similar crises that erupted later: according to the most widely held theory, it was a typical speculative bubble as prices kept rising, fueled by the self-fulfilled prophecy that they would rise even more in the future.[12] The bubble burst when some plant diseases destroyed the bulbs of the most expensive tulip varieties, spreading havoc in the market.

Another factor of skyrocketing prices was considered to be the monetary over-expansion of the time caused by the free coinage policy of the Bank of Amsterdam. Tulip prices collapsed when the Bank forbade the further issuing of currency by private mints, suddenly shrinking the money supply and, by extension, demand for conspicuous goods such as rare tulips.[13] In yet another interpretation, the rising price of tulips was in effect quite similar to modern derivative markets; see Thomson (2006). High value contracts were made for future bulb deliveries and it was possible to change the contract into a call option by paying only 3 % of the latter's price. This meant that the call option contract carried very little risk, while promising high returns given the expectation of further increases in future tulip prices. The market simply collapsed when the Bank of Amsterdam prohibited call-option contracts. Finally, other interpretations find the sharp price rise and fall as a mere symptom of initially expensive fashionable goods, which subsequently got normal after production increased; see Neal and Weidenmier (2002).

6.4.2 The Monetary Crisis in England, 1690

Economic thinkers in seventeenth century England pondered on the issue of credit and interest rates, in an attempt to explain—and imitate—Holland's enviable domination in international trade. In contrast to the highly sophisticated Dutch

[12] According to Garber's classic analysis on famous first bubbles (1990).

[13] In 1637, the metal stock of the Bank of Amsterdam expanded by 52 % over the previous year. Expansion was halted and in 1638 the stock fell by -1 %; data taken from van Dillen (1964).

credit markets, money availability in England was advanced by crude methods such as the widespread practice of coin clipping, causing inflation and undermining trade.

To deal with inflation, England issued new silver shillings, having the same nominal value as the clipped ones, thus the currency effectively appreciated and prices dropped. The new silver shillings quickly disappeared from circulation, prompting Gresham to formulate his famous law that "bad money drives out good money" as the latter is hoarded. In the meanwhile, the real value of loans became higher, ruining many debtors and allowing lenders to amass wealth.

According to economist Josiah Child (1630–1699), the solution was to cut interest rates from 6 to 4 %, in order to stimulate credit and support production and commerce, but Locke dismissed this idea. Based on the premise of price stability (as prices should be pegged to the value of labor incorporated in produced goods), he calculated the quantity of money required for the exchange function of the economy and argued that this quantity should remain stable.[14] The ensuing debate on how to protect the economy from coinage debasement and inflation led to the establishment of the Bank of England in 1691, the first proper central bank in economic history.

In addition to its commercial activities, the Bank was also tasked with lending to the state and, to safeguard the stability of the new currency, adopted in 1717 the rule of convertibility of currency into gold; this rule lasted until the twentieth century, with only brief suspensions during war periods. The UK Gold Standard would become the second longest-lasting monetary reform in history, the first being the reform of Constantine the Great in the Eastern Roman Empire as has been examined in Chap. 4. Following these developments, the English currency was in great demand internationally with London robustly emerging as a global financial centre.

6.4.3 The Financial Crisis in England, 1720

The 'South Sea Company' was founded in 1711 by England in order to facilitate credit and absorb part of the mounting sovereign debt. In exchange, it was granted the monopoly of trade with South America and the public quickly went into frenzy on the expectation of exorbitant profits. At first, company stock prices skyrocketed; the propensity for speculation was so immense that it led to an unprecedented wave of joint-stock 'bubbles'. When actual profitability of the company proved to be much below the initial promises, the market collapsed and a long trail of retributions against company proprietors and complicit politicians began, while

[14] The dispute between Child and Locke has many common features with the conflict between competing economic theories trying to explain the causes and evaluate the measures taken to deal with the Great Depression of the 1930s, as we will see in Chap. 9.

bubble-stocks were banned from trade in drones.[15] But the South Sea company was too big to fail, and Parliament decided that the Bank of England would take on part of investors' losses, by issuing fixed-rate bonds with very long maturities.[16]

6.4.4 The Financial Crisis in France, 1721

In France, public debt was bloated by war-spending and investors lost confidence in the currency, massively turning to gold and silver coins for safety. Desperate for cash, the government created the so-called *Compagnie dell' Occitende* which held the monopoly in the—then French—colony of Louisiana and invited the public to participate in the issue. In 1719 the venture professed to be involved in irrigation works in the Mississippi, while the government granted it with the exclusive privilege of trading to the East Indies, China and the South Seas.

The scheme dazzled the public, stock prices skyrocketed and further subscriptions followed. To enable an even larger number of investors to participate, the Royal Bank—though holding 25 % of company's shares—was instructed to provide financing thus further fuelling stock prices. But when the Company's profitability plummeted and shares collapsed, the same happened with the Royal Bank which had used its own gold reserves to buy shares on the expectation of a huge windfall. The "*Mississippi bubble*" had severe consequences for the real economy, as it led to the collapse of the entire financial system and took many years for the trust to be restored in the French banking system.

6.4.5 Alternative Responses to the Crises

Regardless of their obvious similarities the financial crises in England and France were dealt with differently by the respective governments and impacted very different consequences for their economies.[17] England had a Central Bank that was focusing on maintaining the stability of currency and never got involved with the South Sea Company. When the stock collapsed gold reserves had remained intact, thus boosting the Bank's credibility and making investors eager to accept its bonds as a partial compensation for their stock losses. In contrast, the Royal Bank of France funded its participation in the Mississippi Company by selling its own gold reserves and going down with it.

[15] Mackay, *Extraordinary popular delusions* (2003, pp. 60–63) lists 86 companies that were banned by the Court as bubbles.

[16] Newton himself lost a fortune after the collapse of the company, commenting on the recklessness of investors with his famous phrase: "I can calculate the motion of heavenly bodies, but not the madness of people".

[17] Mackay (*op. cit.*, p. 45) asserts that though "... the madness ... infected the people of England at the same time and under very similar circumstances, [...] thanks to the energies and good sense of a constitutional government, was attended with results far less disastrous than those which were seen in France".

The comparison of the two crises showcases the problems that arose in 2008. An important cause for the recent global crisis concerns the fact that, during the 1990s, banks were allowed to get involved both in retail and investment banking, the latter often including high-risk financial products. When the market value of such products vanished, banks suffered major losses and were threatened with collapse. Today, one of the major issues concerning the efficiency of regulation is forcing banks to separate their investment and credit arms, that is the old British prudence to prevail versus to the then French spree. In a twist of history, the consequences of the 2008 crisis were again asymmetric for UK and France, but this time the other way around. The British banks ended up being heavily more exposed abroad, as external positions of amounted USD 7,324 bn or 270 % of GDP versus USD 3,088 bn by the French banks or 105 % of GDP.[18]

6.5 Repression as Crisis: Slavery and Convenient Theories

6.5.1 Slavery in the Antiquity

In ancient Greece, all laborious chores and activities were performed either by economic immigrants (*metics*) who came to Greek cities from dependent areas or by slaves captured in times of war. Slaveholding was usually individual and most slaves were domestic. They were employed to carry out domestic chores, at farms and workshops, but could be found also in public services as supervisors and surveyors. The *metics* were involved in trade, manufacturing and lending while they often took the responsibility of managing the house, the farm or the business, and could act as assignees in trade transactions. In Rome, this right was also extended to slaves.

When the demand for labor was high—for example in mining, ship-building or the transportation of supplies during wartime—the scale of slaveholding became massive and conquests ensured the regular supply of slaves by forced migration of conquered populations. There were no indigenous slaves, although by the time of Solon many freemen would become enslaved for a specific period, having placed themselves as collateral for loans.

There is no conclusive evidence on the ratio between slaves and free citizens. According to some writers, slaves in Athens corresponded to one-third of total population. However, if metics are included, the ratio changes dramatically: there were 30–40,000 free citizens versus 80–100,000 people in obligatorily employment.[19] In Sparta, each free citizen was assigned seven helots.[20] Most of them were coming from neighboring areas, and forced to cultivate the lands of the sovereign city in a way similar to serfdom in Middle Ages.

[18] Data are from the Bank of International Settlements (2008, Table 2A).

[19] Anderson, *Passages from Antiquity to Feudalism* (1974, p. 38).

[20] According to Herodotus, *Histories*, Book 9, 10.

The philosophers of the time, including Plato and Aristotle, adjusted their teaching to the existing reality and justified slavery, departing from the prerequisite of liberty as was the case in other aspects of their inquiries. Stoic philosophers were more progressive, and argued that all men are equal and should enjoy the same right to freedom. Even the Stoics, however, accepted slavery as unavoidable in some cases, claiming that being a slave did not contradict self determination.

Later in Rome, Stoic philosophy positively influenced some emperors, and slaves gradually acquired the right to cultivate the land, sue abusive owners in courts and buy their liberation using their lifetime savings; they were also considered free by default if abandoned by their master. Roman law allowed slaves to have property; this property could be transferred when they were resold and handed back when set free. Nevertheless, slaves revolted in 73 BC under the leadership of Spartacus, in history's most organized rebellion against extreme exploitation.[21]

Early Christian teaching also tolerated slavery. Despite preaching that all men are equal before God, the Apostles never explicitly denounced slavery nor asked slaves to revolt against their masters.[22] This contradiction is stronger in the writings of Saint Paul: on the one hand he acknowledges that freedom is more useful to a man than slavery[23] and on the other[24] he advises that it should be accepted, arguing that slaves should be loyal to their masters, all the more so if the latter are also Christians.

The Holy Hierarchs were keener to condemn slavery, arguing that the creator "did not make man a slave, but a free person" (John Chrysostom). Nevertheless, the bondage of captives is legalized by Justinian, though there are limitations on the number of slaves each master is allowed to have and rules are set against abusive behavior. Continuing the policy adopted by the Edict of Caracalla in 212 AD, freed slaves automatically become citizens of the Empire.

Even more recent thinkers were willing to justify ancient slavery. Hegel, for example, viewed slavery as a natural outcome of his theory of historical stages. While he considers freedom to be the ultimate good of man and a prerequisite for the advancement of society, he also qualifies this principle by arguing that men should be worthy of their freedom only if they want to be free.[25] As Marcuse notes, Hegel's analysis of domination and enslavement not only justifies the primacy of power in social strife, but also provides the dialectic of this process.[26] Indeed he

[21] Smaller scale revolts had also taken place before in Etruria and Sicily. See Strauss, *The Spartacus War* (2009).

[22] "There is neither Jew nor Greek, slave nor free, male nor female", New Testament, *Galatians*, 3: 28.

[23] "Art thou called being a servant? care not for it; but if thou mayest be made free, use it rather", New Testament, *Corinthians* 7:21.

[24] "Let as many servants as are under the yoke count their own masters worthy of all honor", New Testament, *Timothy* 6:2.

[25] Hegel, *Philosophy of History*, Introduction (1956).

[26] Marcuse, *From Luther to Popper* (1983, p. 110).

accuses Hegel that his theory of historical stages legitimizes enslavement as the real—and therefore necessary—demonstration of domination.

6.5.2 Slavery in the Modern Era

Later, in Islam, enslavement of the *infidels* is justified by the Quran and slave markets are massively supplied by war prisoners, mainly from Africa. Slaves are employed in trade, farming and domestic work, but also in the army (one example is the elite corps of the Janissary).[27] During the Islamic conquests (from the seventh to fourteenth century), slave trade became a multinational enterprise with its epicenter in Africa.[28]

Slave-trader companies were established and evolved into giant enterprises after the Pope legitimized the enslavement of unbelievers and gave the monopoly to the Portuguese King, who started the importation of slaves to Europe to be used as agricultural laborers. Spain and other colonial powers also became involved in the slave trade, expanding its supplies to South Africa and Asia.

During the sixteenth century, the first slaves arrived in America, to work as a massive, cheap labour force in the large-scale plantations. Cotton, tobacco, coffee and sugar production rose to unprecedented levels, changing the landscape in global trade.[29] Contemporary thinkers reached conclusions about slavery that were quite different than what we would have expected today. In economic analysis, the slave trade was considered necessary for production, while in political theory, it was viewed as a legitimate institution, compatible with the right of expansion by colonial powers and the concomitant exploitation of the new lands.

Even Voltaire, a major figure of the Enlightenment, was unable to see beyond of his time and never condemned slavery. Forgetting his own teachings about liberty, he argues that one who is voluntarily subordinated to a master is destined to be a slave. Justifying practices of slavery, including the enslavement of children, he argued that "those selling their children are more condemnable than those buying them".[30] In the same spirit one of the greatest advocates of democracy, Alexis de Tocqueville, writes in 1835 that "[t]he most formidable of all the ills that threaten the future of the Union arises from the presence of a black population upon its

---

[27] In some activities, however, there was no clear-cut distinction, however, between masters and slaves. In the case of concubines, it was allowed to legitimize the offspring of the master, while the woman slave could not be resold.

[28] The slave trade continued in some areas of Africa even during late twentieth century; according to UN estimates in 1980, around 90,000 Mauritanians were enslaved to Berber chieftains. In Sudan civil war prisoners were massively sold to arms dealers even in early twentyfirst century, while the ongoing slave trade in Mauritania was denounced by the UN as recently as 2009; source *Reuters*, 3/11/2009.

[29] In his seminal book *Capitalism and Slavery*, Eric Williams (1996) mentions that slave trade profits were used to accumulate capital during the Industrial Revolution.

[30] Voltaire, *Essay on the Manners and Spirit of Nations* (1979, p. 551).

territory". Blacks are dismissed because "[their] physiognomy is to our eyes hideous, his understanding weak, his tastes low; and we are almost inclined to look upon him as a being intermediate between man and the brutes".[31]

Only after the US became independent did the long process of improving their living conditions commence, culminating in the abolition of slavery in 1865. In Britain slave trade was abolished earlier in 1806, though the British colonies kept it until 1833. The abolition of slave trade was not the result of any large-scale slave revolt, though some acts of resistance did occur in various places.[32] It was rather an economic trend that gradually undermined the advantages of slavery: as production in Europe and America was becoming technically more demanding the pressure for specialization and expansion of the labour force was also increasing. The free labour force was becoming more specialized, reliable, adaptable to new conditions and, ultimately, more productive,[33] while the cost to buy and keep slaves was increasingly heavier.

As capitalism prevailed in the nineteenth century, the exploitation of slaves was no longer needed to ensure profitability. The mass society that emerged required individualism rather than blind obedience so that mass production and consumption brought about by the Industrial Revolution could be effectively carried out.

The slavery debates are still raging on to date.[34] Fogel and Engermann (1974) presented productivity estimates in the American South during the antebellum period and concluded that the system of slavery in America was economically efficient; returns were high and the rate of exploitation rather low.[35] In Fogel's interpretation, emancipation was not driven by economic forces but was the outcome of a moral awakening that placed egalitarianism as the centre-piece of the value system in the US.[36] Their conclusions attracted a lot of criticism both on the grounds that were providing an economic justification to the slave trade and regarding their selective use of data sources.[37]

---

[31] Tocqueville, *Democracy in America* (2003, I, Chap. 18).

[32] In Haiti, the slaves revolted and finally succeeded in establishing an independent state in the late 18th century. Another state established by slaves was Liberia, when the US agreed in 1847 to release them under the condition that they depart and return to Africa. Today many Liberians have American names and the state's capital Monrovia was named after James Monroe, President of the US at that time.

[33] See Marx, *The Revolution of 1848* (1973, p. 297).

[34] For an account of these theoretical confrontations see Fogel, *The Slavery Debates* (2003, p. 28).

[35] Fogel and Engerman, *Time on the Cross: The Economics of American Negro Slavery* (1974).

[36] Fogel, *The Fourth Great Awakening and the Future of Egalitarianism* (2000).

[37] For example see Gutma, *Slavery and the Numbers Game: A Critique of Time on the Cross* (2003).

References

Anderson P (1974) Passages from antiquity to feudalism. Verso, London
Bank of International Settlements (2008) Quarterly Review, September
de Tocqueville A (2003) Democracy in America. Penguin, London
Engels F (1947) Anti-Dühring (Translation: Emile Burns, from the 1894 Stuttgart third edition). Available at http://www.marxists.org/archive/marx/works/1877-ad/index.htm
Fogel R (2000) The fourth great awakening and the future of egalitarianism. The University of Chicago Press, Chicago
Fogel R (2003) The slavery debates: a retrospective 1952–1990. Louisiana State University Press, Baton Rouge
Fogel R, Engerman S (1974) Time on the cross: the economics of American Negro slavery. Little, Brown and Company, Boston
Garber P (1990) Famous first bubbles. J Econ Perspect 4(2):35–54
Gutma H (2003) Slavery and the numbers game: a critique of time on the cross. University of Illinois Press, Urbana
Hegel G (1956) The philosophy of history. Dover, New York
Hobbes T (1982) Leviathan. Penguin, London
Kant I (1977) An answer to the question: what is enlightenment? In: Reiss H (ed) Political writings. Cambridge University Press, Cambridge
Kitromilides P (1998) Political thinkers of the modern era: biographies and interpretations. Poreia, Athens (in Greek)
Locke J (1690) Second treatise on civil government. Available at: http://www.constitution.org/jl/2ndtreat.htm
Locke J (1977). In: Yolton J (ed) The Locke reader. Cambridge University Press, Cambridge
Mackay C (2003) Extraordinary popular delusions. Dover, New York
Marcuse H (1983) From Luther to Popper. Verso, London
Marx K (1973) The revolutions of 1848. Penguin, London
Monroe AE (2006) Early economic thought: selected writings from Aristotle to Hume. Dover, New York
Neal L, Weidenmier M (2002) Crises in the global economy from tulips to today: contagion and consequences. NBER Working Papers 9147. National Bureau of Economic Research
Rousseau JJ (1981) Reveries of the solitary walker. Penguin, London (Trans. P. France)
Shrady N (2009) The last day: wrath, ruin, and reason in the great Lisbon earthquake of 1755. Penguin, New York
Strauss B (2009) The Spartacus war. Simon & Schuster, New York
Thompson EA (2006) The tulipmania: fact or artifact. Public Choice 130:99–114
van Dillen JG (1964) History of the principal public banks. A. M. Kelley, New York
Voltaire (1979) Essay on the manners and spirit of nations. In: Redman BR (ed) The portable Voltaire. Penguin, London
Williams E (1996) Capitalism and slavery. Wiley, New York

The Industrial Revolution and the Foundation of Classical Economics

7

Abstract

On the eve of the industrial revolution, new economic theories were sought to displace mercantilism and Physiocracy, and instead promote savings and investment as the main drives of economic growth. A combination of technological advance, market incentives and better institutions made England the hotbed of the new age, leaving behind wealthier—yet dysfunctional and repressive—France. Adam Smith provided a theory on how markets can ensure unhindered progress for the mankind, laying the foundations of modern economic theory. Subsequently, Malthus warned on the limited availability of resources and the risk of a society collapsing if population expansion is not checked. Ricardo reaffirmed the potential of continuing growth if capital accumulation is freed from high land rents and trade is liberalized so that each nation is specialized in its competitive advantage.

7.1 The Ascent of Industry and the New Economic Theories

Three events in the late eighteenth century revolutionized the economic, sociopolitical and geopolitical landscape worldwide: the Industrial Revolution, the French Revolution and the Declaration of American Independence. The new horizons opened up by technological advances were combined with the unlimited potential of trade and the search for a better life by previously oppressed populations, alarming the rulers and posing new questions for economic thinkers. As scale of production and freedom of movement were advancing at a quick pace, the new paradox they focused on was how a free-willing individual can act in a liberated society and improve his position without anyone calling the shots in his stead. Answers differed across countries.

N. Christodoulakis, *How Crises Shaped Economic Ideas and Policies*,
DOI 10.1007/978-3-319-16871-5_7

7.1.1 Social Turmoil and Balances

Addressing this issue, the Dutch thinker Bernard Mandeville (1670–1733) argued that in a society where rules exist and freedom is exercised under specific limitations, an individual can serve the general good even if solely motivated by self-interest. Mandeville's famous phrase "private vices and public virtues" is reminiscing of Machiavelli's recognition of diverging self-interests and echoes Hobbes's views on how their excesses can be harnessed. Mandeville's innovative conclusion was that the public good is served not by the diktats of an almighty state, but through the operation of a free market. The government should not interfere, except to promote employment so that all citizens have opportunities to exercise their freedom and gain prosperity.

In France, the libertarian movement was philosophically and politically expressed by the Enlightenment. Though not a coherent ideological movement, it favored progress, technical discoveries and freedom, but stopped short of formulating a comprehensive economic theory. The most important Enlightenment thinkers were Montesquieu, who articulated the theory of separation of state powers, Diderot, an *encyclopaedist* who studied the structure and dissemination of knowledge, Voltaire who stood for the freedom of expression and political discourse, and Rousseau who fervently opposed the transition from the natural state of subsistence to an economy relying on increased productivity and growth, as we have already seen in Chap. 6.[1]

Despite the lack of a coherent economic thought, the period just prior to the French Revolution of 1789 witnessed some important developments in rationalizing the conduct of economic policy. For example, Comptroller General A. R. J. Turgot abolished taxation on grain, introduced a single tax and nationalized the post offices. Very soon, however, he was sacked by the King as too radical for the interests of the nobility.

7.1.2 England on the Rise

In England, economic thought was advancing at a much quicker pace. The emerging power of industrialists and professionals had started to evolve into a social class, which demanded recognition, influence and power. The new class faced the rivalry both of the landed nobility that controlled the agricultural sector, and of merchants who undermined domestic production through massive imports. Economic theory was gradually formulated as a new way of thinking, distant from both mercantilism and physiocracy, and compatible with the upcoming industrialization. It was the era of classical economic theory.

From a technical point of view, the Industrial Revolution was caused by the waves of innovation in the production process. Technological advances

[1] Colletti, *From Rousseau to Lenin: Studies in Ideology and Society* (1974, p. 155).

dramatically increased productivity, cut down production costs and opened up new markets for English products in Europe and the rest of the world. In the mid-eighteenth century, technical knowledge thrived and production increased rapidly. People abandoned the land to find work in the cities, and, despite the huge increase in labour supply, wage levels not only escaped downward pressures, but actually rose in real terms. From being one of the poorest nations in Western Europe in 1700, by the late eighteenth century England had become one of the wealthiest with national income higher by 43 % in per capita terms. Over the same period, income per capita increased by a moderate 23 % in the Netherlands and declined by −8 % in Northern Italy.[2]

7.1.3 The Gap with the East

But technical innovation is only part of the picture and, on its own, cannot explain two crucial questions: the first is why the Industrial Revolution occurred in Europe and not East Asia and more specifically in China, also technologically advanced at that time. An explanation might be that China had no need for leaps in technical innovation and production, because—as happened in ancient Rome as well—it had access to a huge and cheap agrarian labour force. People were not allowed to move in order to find work but had to stay close to their hometowns, so that the Empire could tightly control both population and crops.[3]

The differences between China and Europe have always attracted the interest of Western thinkers. According to Hegel, the differences between East and West are explained by a mixture of rationalism and necessity. The geographical division is not a chance event, but represents a fundamental rupture in History–a divide in the world of ideas that takes shape in the material world. The West is identified with the elements of change and advance, while the East is only partially considered the embodiment of the spirit of freedom. As such, a foreign conquest cannot–according to Hegel–downgrade the character of the empires in China and India, "...because in the state of non-freedom they got entrapped, they cannot fall any lower".[4]

In the nineteenth century, John Stuart Mill had argued that China was doomed to remain stagnant, despite its important advances because seeking uniformity in all manifestations of life permeated society and suffocated innovative thinking; see Mill (1981). Similarly, Voltaire had identified several differences between Europe and

---

[2] Derived by using data from Bolt and van Zanden (2013), *Table 2 GDP per capita in various parts of the world, 1348–1800*.

[3] A similar system of attachment to a specific location can be found even in modern China. The infamous *hukou* system essentially forbids the Chinese to move permanently from the country to urban centers. If a Chinese citizen wants to start a family, for example, he has to return to his place of birth, since many benefits, including the right of his children to go to school, are only available there. The number of Chinese citizens falling into this category is around 130 million; for a description see Cohen, *La Prospérité du vice: Une introduction (inquiète) à l'économie* (2009).

[4] See Hindes and Hirst, *Pre-capitalist Modes of Production* (1979, p. 204).

China, from religion and politics to its system of governance and nutrition. He even makes special reference to the large number of eunuchs in Asian courts and highlights the great differences between Europe and the East in the treatment of women. By way of illustration he says that no woman had ever reigned in the East, whereas in Europe women were entitled to the throne in many nations.[5]

The debate continues to date and alternative theories are suggested. Clark (2007) evokes the "scale argument" and explains that

> Europe made this leap because it had coal reserves readily accessible to its population centres [and] in addition it had the massive, largely empty land area of the Americas . . .to lift for a time the ecological constraint.[6]

Findlay and O'Rourke (2007) argue that China never lacked the space to expand its activities and heavily criticize the explanations relying on the notion of "an unchanging East" by providing manifestation of widespread trade and market activities as well as masterly achievements in defense technology and public infrastructures. Alternatively, they suggest that the rise of a unique and competitive system of European statehood since the seventeenth century unleashed a process of antagonism as well as a systematic search for survival and supremacy. Ultimately it became the attractor of technological innovation and mass mobilization of resources to satisfy the increasing scale of production.[7]

7.2 Industrial Revolution: A Tale of Two Countries

But then a second, and more difficult, question comes forth: if the Industrial Revolution sprang out of the European state-system, why then occurred in England and not in France?

7.2.1 Could It Have Been France?

If someone in the early eighteenth century had posed the question on which country would be more likely to give birth to the industrial era, most analysts would probably point to France since it was ahead of England in many crucial aspects, including the following:

(i) *Large population and domestic market*: France had a population of 21.5 million in 1700, by far the most populous nation in Europe and two and half times larger than the combined population of 8.6 million in Britain and

[5] Voltaire, *Essay on the Manners and Spirit of Nations* (1979, p. 552). He was actually wrong as Empress Wu Zeitan did reign during 690–705 AD overpowering the weak Emperor Tainzog.

[6] Clark, *A Farewell to Alms: A Brief Economic History of the World* (2007, p. 260).

[7] Findlay and O'Rourke, *Power and Plenty: Trade, War, and the World Economy in the Second Millennium* (2007, p. 359).

Ireland.[8] France's much larger domestic market was ideal for economies of scale and its labor force could sufficiently meet the needs of an expanding industrial sector.

(ii) *Advantageous geographic location*: France could communicate with the rest of continental Europe and Mediterranean countries without the risks and costs of sea transport that England was facing. In addition, France's neighbors were all well developed and rich states—a definite advantage for French exports.

(iii) *Superiority in science and the arts*: France witnessed significant developments in science—especially since Descartes' contribution to mathematics—whereas England was clearly a laggard, showing progress only after Newton's advances. Besides natural sciences, French arts were also booming with great achievements in theatre, opera, music and literature.

(iv) *Advanced way of life*: Per capita consumption in France was qualitatively and quantitatively superior to England's.[9] The French state was also richer. After the Mississippi bubble burst earlier that century, the French crown had amassed bullion and coin reserves which far surpassed the wealth of the British Treasury.

7.2.2 Why It Was England

Why then did France not take advantage of these auspicious conditions—material, human and geographical—to lead in the industrial era? According to modern theories of growth other factors, despite being less obvious, may play a crucial role in shaping economic progress. These include the quality of institutions, property and individual rights, mass education, religion and openness. These can be catalysts for progress, allowing a state to overcome obstacles and disadvantages in other sectors. Furthermore, economic deprivation and increased risk in business activities, provided they are not too excessive, can actually act as a motivation rather than an impediment to growth. Some sectors where England had a critical advantage over France were the following:

(a) *Greater economic stability*: In the beginning of the fourteenth century, England had gradually established a more effective system of fiscal management, ensuring higher tax compliance of her subjects and better resource allocation. The opposite was the case in France. The crown, in order to raise more revenues, imposed domestic tariffs on the transportation of goods and steadily increased the tax burden on producers. The collection of taxes was

[8] Data are from Maddison historical population series.

[9] One hundred years later, in 1870, British GDP per capita had more than doubled compared to its levels at the beginning of eighteenth century. It was also 1.7 times larger than French per capita GDP and 16 % higher than Holland's. Data of GDP per capita are from Maddison historical series.

commissioned to private individuals who, in order to maximize their own profits, further squeezed taxpayers.

Following the crisis of 1690, England had made great progress in putting her economy on a stable footing and restoring the credibility of its currency. The establishment of the Gold system, combined with the control of American goldmines, became the anchor of stability in the international monetary system and made London the most important financial center in the world. In France, the banking system had collapsed after the Mississippi crisis, and took many decades to recover. During the eighteenth century, England also introduced new methods to record public spending, and thanks to conducting a regular census the government was able to study the main characteristics of population growth. France on the other hand did not progress much after Quesnay's *Tableau*.

(b) *Political representation*: By 1700 the parliamentary system in England had become much more progressive compared to the authoritarian regime in France. Following the beheading of King Charles (1684), the English Crown ceded important powers to Parliament, while voting rights and electability were expanded. Parliamentary tradition absorbed many social tensions well before these boiled over and led to revolution and dramatic upheavals. In France, it was as late as 1787 that Louis XVI ceded some powers to the General Assembly in a last ditch attempt to thwart the revolts caused by the famine. But it was too little and too late so the measure backfired: public anger was skillfully channeled through the Assembly and quickly morphed into a revolution.[10]

(c) *Greater influence on emerging economies*: France might have been the heart of the "old world" in Europe, but her influence was steadily decreasing in the "new world" of America. After the Treaty of Utrecht (1713) it lost many of its North America territories to England, while French attempts to gain a foothold in the Southern Cone (where Spain and Portugal were the undisputable authority), ended abruptly. The French retained their stronghold in Africa, but even there they were challenged by the strong English presence in the centre of the Black continent and the growing Dutch influence in the southern part.

In late eighteenth century, the British metropolis was receiving massive inflows of gold and silver from its American colonies. London used the proceeds to build an army and fleet unrivaled anywhere in the world. With the Industrial Revolution, colonies could call on an inexhaustible source of cheap labor (and even cheaper slavery), while also providing all the raw materials and intermediate goods needed in England. Colonies also proved to be huge market outlets for domestically manufactured products, while France retained only a niece of international trade.

[10] According to Lefebvre, *French Revolution* (1962, p. 137), the combination of the two fostered a rebellious mindset that eventually led to the French Revolution.

(d) *Employment and market conditions*: In England, the peasants and their families who worked the land were still dependent on landlords, but the latter had no legal jurisdiction over them, nor could they forbid them from moving. In France, peasants were forbidden to move and, moreover, the *corvée* regime obliged them to spend part of each year working without pay. It is hardly surprising that the number of people who were destitute rose inexorably and peasant revolts were frequent. According to Lefebvre, "[i]n France, one in five citizens was poor, and their numbers dramatically increased after the economic crisis. In addition, France had a non-existent welfare system".[11] The combination of these two factors fostered a rebellious mindset that eventually led to the French Revolution.

In contrast, a compulsory "poor tax" was imposed at the parish level in England to support the most deprived. The English care system had been considerably expanded and 154 hospitals were established in the period 1700–1825, funded mainly by private donations.[12] Another progressive feature in England was the establishment of a single internal market and the abolition of barriers and tolls which had existed in Middle Ages. In France, by contrast, toll stations were abundant, even on the main rivers. The result was that a product transported on the Seine would see its price increase by as much as 50 %. Germany had a similar model, with dozens of toll stations along the Rhine and the Elbe.[13] This made the transportation of merchandise an incredibly complex, time-consuming and expensive venture.

(e) *Technique and Technology*: In the eighteenth century, England had already instituted mass education. By 1730, approximately 70 % of men and 25 % of women had received basic education[14]; the respective figures for France were a fraction. Though natural sciences were thriving in France, knowledge remained a privilege of the few. Since most schools belonged to the Church, the dissemination of technical and scientific knowledge was limited and under theological control. For any thinker to succeed and his theories to become known, he needed the support of the royal court and the sponsorship of rich merchants. Manufacturers who might have been the actual beneficiaries of scientific and technical progress were not part of the picture.

Technological progress needs incentives—and motivation was abundant in England but completely lacking in France. In England, public interest in patents and inventions was feverish, with open speeches and periodicals spreading the

---

[11] Lefebvre, *op.cit.*, p. 66.

[12] Trevelyan, *English Social History* (1946, p. 346).

[13] For a description see Heibroner and Mildberg, *The making of Economic Society* (2008, Chap. 3).

[14] Data are from Broadberry, *Recent developments in the theory of very long run growth: a historical appraisal* (2007, Graph 2).

news, thus promoting both the work and the social acclaim of inventors.[15] The steam engine was invented in England, precisely thanks to an environment of constant pressure to mechanize the textile industry and decrease costs, so that English manufacturers would be able to compete with the Dutch.

The French economy on the other hand had not adopted innovation to boost production in the manufacturing process and was relying on demographic growth and high prices to increase turnover.[16] Still under physiocratic influences, France dismissed the value of technical progress and understated the contribution of manufacturing to social prosperity. With no social acclaim for innovators, no moral legitimization of progress and lacking the pressure of competition, the industrial revolution could ever have happened in France. Though many of the preconditions for technical progress existed, France was able to exploit some of them only after the Revolution.[17]

(f) *The Seven Years War (1756–1763)*: This involved most of the great powers of the time, but was mainly driven by the rivalry between England and France. At stake was the control of colonies in America and Asia, especially in areas producing important commodities for the European economies (sugar, tea, cotton, etc.). France was allied with Spain as both countries were ruled by the House of the Bourbons, while Austria and Prussia took the side of England. France was in no position to defend her overseas colonies and, as a consequence, suffered a major defeat due to the supremacy of the British navy. Under the Treaty of Paris (1763), France was forced to cede Quebec and Louisiana in America, as well as Bengal and Hyderabad in India, while Spain ceded Florida.

England came out of the Seven-Years War as the undoubted international superpower of the time, as well as dominating in global trade. Soon, however, the declaration of American Independence (1776) would bring its colonial domination in the New World to an end. The defeated nations of the Seven Years War (France, Spain and Holland) took revenge by equipping the American revolutionaries in their war against the English crown. France and Spain actively participated both in the American war of Independence and in the institutional design of the independent United States. But this support backfired as it drained domestic resources and aggravated social discontent.

[15] An example was the Gentlemen's Magazine, which, according to since 1729 had been a publication dedicated to informing its readers about new inventions, as commented by Heilbroner and Midberg, *op. cit.*, (p. 142).

[16] Lefebvre, *op. cit.* (1962, p. 66).

[17] Even after the French Revolution, science was not considered a major component of social progress. During the Reign of Terror, the great chemist Lavoisier was tried as a traitor (despite having participated in the revolution). When he appealed to the judge to spare his life in order to continue with a very important experiment, the judge is said to have replied that "the Republic needs neither scientists nor chemists". Lavoisier was guillotined in 1794.

(g) *Religious tolerance and social dynamics*: The clash between England and the Vatican in 1487 resulted in a greater religious tolerance, with Protestantism eventually becoming the dominant doctrine. In France, on the other hand, the protestant Huguenots were relentlessly persecuted. After the massacre on St. Bartholomew's Day in 1592 they were forced to flee the country, seeking refuge mainly in England and the newly discovered America. Since many of them were scientists, merchants and members of the rising bourgeoisie, France suffered a tremendous loss of specialized human capital—to the benefit of protestant nations. More religious persecutions followed on a massive scale, suppressing the freedom of thought and the spirit of inquiry in many sectors, and transforming France into an illiberal and oppressive regime.

J. S. Mill noted that there was another difference between the two societies: in Britain there was no monolithic uniformity of goals and behaviors, leading to a dynamic society, whereas in France society was static and stagnant. According to many historians, the class barriers in England never became impenetrable walls, separating the nobles from the members of the bourgeois class. Nobles had no visible prejudice against business people who were also trying to climb the social scale.[18]

(h) *Structure of the agricultural sector*: In England, peasants were tenants who cultivated the land under long-term leases and the landlord had no legal right to unilaterally break the contracts. Laws against ejections had existed as early as the fourteenth century. In Scotland, in 1449 a law was passed that secured the leases from successors of any kind, being either new owners or heirs. Leases were often for life, and rents were pre-agreed each year. This gave farmers the motivation to increase production and invest the surplus in improving productivity, since the beneficiaries would be themselves too.

France was a laggard in this aspect as well. Lease terms were limited to 9 years, while taxes were proportional and based on the production estimates made by the private tax collectors acting on behalf of the sovereign. To avoid paying part of their produce, renters hoarded production without investing in the improvement of their land. As Adam Smith argued with regard to France, "avarice and injustice are always shortsighted, and they did not foresee how much this regulation must obstruct improvement, and thereby hurt in the long-run the real interest of the landlord".[19]

[18] Lefebvre, *op. cit.*, (1962, p. 67).

[19] Smith, *The Wealth of Nations* (1986, Book III, Chap. 2, The discouragement of agriculture, p. 492).

7.3 Adam Smith: The Economics of Optimism

Adam Smith (1723–1790) was born and educated during the period of the Scottish Enlightenment; hence he was influenced by the views regarding the importance of savings to be critical in securing a nation's well-being and the balance of trade in preserving the purchasing value of money. He vouched to interpret the new economic environment of the Industrial Revolution and, in attempting to do so, was the first to author a comprehensive treatise on economics with firm philosophical foundations. His work analyzes real phenomena, measures their consequences and explains their practical significance, while also proposing a system of rules for the advancement of general prosperity.

The unresolved issue that he inherited from his teacher, Francis Hutcheson, was how can society collectively progress if its members are motivated by self interest alone? Mandeville's answer had been more intuitive than analytical, and this did not satisfy the philosophical standards of Smith. The question was urgent, as the wave of urbanization was rampant, new production techniques increased productivity at unprecedented levels, and education was spreading to the masses. Mercantilism had nothing to say about how to handle domestic production, while the physiocrats—Smith, as well as many other representatives of the Scottish Enlightenment, felt nothing but contempt for their doctrines—believed that the only concern of the state is to support the agricultural sector. Smith accepted Hutcheson's view that mutual sympathy and support help society, but did not believe in imposing limits on individual self-interest so as to make behavior more altruistic. He was looking for a more resilient and robust process of balancing individualism with the general good.

In his *Theory of Moral Sentiments*, written in 1759, Smith paves the way for his great contribution to economic theories. He offered an answer to the age-old issue of how collective harmony can be achieved even though society consists of disorderly and individualistic members. He argues that, a just and organized society based on the rule of law can accommodate individual self-interest and utility. Smith uses insights of previous thinkers, and combines them to give a coherent account of this balancing mechanism. In his best-known work and crowning achievement, *The Wealth of Nations* written in 1776, he gives an account of the factors contributing to economic progress and individual prosperity. Ever faithful to Locke's empiricism, Smith begins his work by defining all the value-producing factors: labor, capital stock and land, paid for by means of wages, profits and rents respectively.

Central in Smith's analysis is how crucial the division of labour is in reducing costs and increasing productivity. Building on the views of the Scottish Enlightenment, he also makes a distinction between productive and unproductive labour: the former adds to the value of the commodity upon which it is bestowed, while the latter does not. This view can be explained in part as a result of his distaste for mercantilism, but is also an argument against luxurious, imported consumption and in favor of savings, as only the latter would allow the growth of domestic industry. For an increase to happen in the capital stock, investment is needed and is financed from foregone consumption.

To prove that an economy ends up in equilibrium, he argues for the existence of an *invisible hand*, which through the market conveys individual goals and personal ambitions to the benefit of society. According to Smith, the stable expansion of the economy and the accumulation of investment open up new opportunities, with the division of labour increasing productivity and the market mechanism balancing desires and motivating further expansion. No crisis could erupt from a gap between supply and demand, thanks to the market's equilibrating mechanism. Economic philosopher Mark Blaug proposes that the invisible hand should not be interpreted solely in a static way that ensures an efficient division of labour during a specific period of time, but rather as a process of economic development, which must be supported by institutions and state policies.[20]

A popular misinterpretation of Smith is to present him as a diehard of *laissez-fair*, though in fact a major part of his *Wealth of Nations* assigns major duties and functions to the Government.[21] Smith was in favor of progressive taxation and thought that the state should be responsible for security, justice and institutions. The functioning of the market should not hinder the provision of public goods; thus Smith supported the deregulation of education so as it could be provided not only by the state but also by private interests to reach as many as possible; other public goods and state services should be funded by tariffs or tolls. Opposing the mercantilist views, he rejected protectionist measures in support of domestic industry. Smith also condemns monopolies and is a fervent proponent of unhindered competition, in order to maximize the mobility of production factors and society's well-being. His is fearful of practices that go against the interests of consumers, and dismisses all privileges bestowed on commercial enterprises. Warning on the danger of harmonized practices he notes that "people of the same trade seldom meet together... but the conversation ends in a conspiracy... in some contrivance to raise prices".[22]

As often is the case with great thinkers, Smith has also many critics. The main criticism against him claims that his work is not original but a compilation of previously expressed views: from Xenophon's division of labour, to Mandeville's contradictory interests, to Galian's price equilibrilism mechanism.[23] Regardless of the merits of this criticism, there can be no doubt that Smith was the one who made a comprehensive evaluation of previous views, turning them into a coherent, interesting and convincing theory. He is therefore rightfully considered the founder of modern economic science.

---

[20] Blaug, *Economic Theory in Retrospect* (1978, p. 63).

[21] For a discussion see Himmelfarb, *The Roads to Modernity* (2008, pp. 53–70).

[22] Smith, *op. cit.* (1986, Book 1, Chap. 10, part 2).

[23] Galiani (1728–1787) published *Della Moneta* (On Money) in 1751 where he described the mechanism of equilibrating demand and supply: When production of goods is raised "...the use of them increased: which use prevented their value from falling as much as their abundance required. And so this interrelation produces the great and very useful effect of equilibrium of the whole... although it results, not from human prudence or virtue, but from the base incentive of sordid gain". Quoted from Monroe, *Early Economic Thought: Selected Writings from Aristotle to Hume* (2006, pp. 297–298).

7.4 Malthus: The Dismal Economics

In the period following the early industrialization, Robert Malthus (1766–1834) emerged in England as a prophet of an ominous future, warning about the diminishing capacity of production to feed an ever-growing population. In contrast to Adam Smith's optimism, Malthus was the theorist of a perennial state of crisis and for this reason economic theory in general has been called the "dismal science". He believed that crises, rather than being mere episodes of chance, were inescapable and fatal. Somehow overlooking the great potential generated by the production revolution, Malthus warned about a likely economic meltdown arguing that:

> It may be expected, indeed, that in civilized and improved countries, the accumulation of capital, the division of labour and the invention of machinery, will extend the bounds of production; but we know from experience, that the effect of these causes . . . are very much less efficient in producing an increase of food.[24]

According to Malthus, while the amount of subsistence increases by arithmetic progression, population grows by geometric progression, inexorably leading to supply shortages. Malthus could not see any solution coming from either higher agricultural productivity brought about by new techniques or by importing food supplies from countries facing no population pressure. His arguments lead to the inevitable conclusion: if the world wants to preserve the capacity to feed its habitants, further population growth should be prevented.

Malthus thought that such self-preserving choices can be reached only by civil and political liberty and education; otherwise it would be imposed by despotism and oppression.[25] Societies should be adopting birth-control policies and, for the same reason, philanthropy and the laws supporting the poor should be abolished. Naturally he sided with landowners, for he believed that they play a key role in producing the food needed to feed society.

Regardless of the revulsion expressed towards Malthus's theories, history seemed to offer a lot of validating cases.[26] Around the turn of ninth century in Central America, the rising population of the Maya was stressed as intensive agricultural production was unable to keep up with demand, causing the collapse of the Empire.[27] A similar Malthusian case was the destruction of the Easter Island civilization.[28] Today, birth control policies widely implemented in India, China and Africa echo what Malthus had suggested two centuries earlier. Even his most

---

[24] Malthus, *An Essay on the Principle of Population* (1985, p. 244).

[25] Malthus *op. cit.*, (p. 252).

[26] Engels, for example, called the Malthusian population theory "the crudest, most barbarous theory that ever existed", (cited from the Introduction to Malthus, *Outlines of a Critique of Political Economy*, p. 51).

[27] For a thorough account on how the interactions between population increase and available resources affected the Maya collapse, see Tainter, *The Collapse of Complex Societies* (1988, pp. 152–178).

[28] For a description of Malthusian collapses in human history see Diamond, *Collapse* (2005).

extreme thesis, that vulnerable individuals should be left unaided for otherwise they have an incentive to remain weak, is not as distant from some modern prescriptions of cutting down welfare entitlements for persons on the dole as a means to motivate them in accepting low paid jobs.

There are, however, many more cases disproving Malthusian pessimism. Productivity in the agricultural sector increased spectacularly, and humanity itself solved many problems of food sufficiency. Though famine crises did not entirely disappear, they were seriously contained. It is perhaps an irony of history that Malthusian crises, which are in essence market failures, rarely happened in societies inspired by the individualistic theories of Adam Smith: in most cases in the twentieth century they were encountered in systems—such as China or Soviet Union—based on the renunciation of markets in the hope of avoiding such shortages.

7.5 Ricardo: The Economics of Accumulation

David Ricardo (1772–1823) became the theorist of the rising business class. His view on rents differs from Smith's, as Ricardo dismissed the notion that land contributes to economic growth. Indeed, he saw landowners as an obstacle to industrialization as the rents paid to landowners exercise a downward pressure to profits and wages, thus hindering further accumulation of capital. According to his theory, the land that is cultivated is always the most fertile and agricultural production continues while it is sufficient to feed the urban populations. As capital accumulates and industrial production expands, demand for labour increases, as does demand for food in the cities. At the same time, agricultural workers leave the countryside and move to cities in order to work in factories. For agricultural production to be able to meet the increased demand, less fertile lands are now utilized. Therefore, landowners ask higher rents for the more fertile areas, thereby diminishing the share of business profits and wages.

Since wages cannot drop below a minimum level necessary for subsistence and reproduction of the work force, it is corporate profits that finally take the hit and businessmen are forced to abandon their investment and expansion plans. We can easily understand why Ricardo dismissed the "Corn Laws", which imposed heavy tariffs on imported foodstuff so as to confer further gains to landowners. The solution, according to Ricardo, was the abolition of tariffs altogether, encouraging the importation of cheaper food. This would allow the English economy to focus on production and exports, compensating for the cost of imports.

As a fervent promoter of international trade, Ricardo formulated the law of competitive advantage which is a central premise of modern economics and described the equilibrating mechanism in external imbalances by arguing that trade deficits and surpluses are eventually balanced by gold flows in reverse. The inflow of gold creates a surplus, which bolsters the value of the currency, leads to domestic price increases and finally leads to home products being less competitive in international markets. The opposite process takes place in the case of a nation with trade deficits—so all in all, the flows of gold ultimately correct the imbalances.

Ricardo was a great analytical thinker and his work is far more comprehensive than either Adam Smith's descriptions or Thomas Malthus's warnings. His often obscure, yet penetrating and insightful writings touch on a multitude of issues, including the effects of taxation, the fiscal consequences of war spending and the theory of value, based on incorporated labour. His work inspired many later economists, from Marx who adopted his labour theory of value, to Marshall who focused on land rent, and from Sraffa to several neoclassical advocates of free trade.

From a social and overall point of view, however, the economist who perhaps would best fit Ricardo's paradigm was J. M. Keynes, exactly a century later. Both were prodigious personalities: deeply involved in politics, well-educated, and with a solid grasp of mathematics. They wrote in an obscure fashion for the average reader of their age and both made fortunes in the stock market.[29] Furthermore, they developed their theories after great wars—the Napoleonic Wars in the case of Ricardo and World War I in Keynes's. Post war, both thought that the Bank of England followed wrong monetary policies: during the Bullion controversy Ricardo condemned the expansionary policy, while in the 1920s Keynes contemptuously called the conversion rate adopted by the Bank as a "relic of barbarism".[30]

References

Blaug M (1978) Economic theory in retrospect. Cambridge University Press, Cambridge

Bolt J, van Zanden JL (2013) The first update of the Maddison project: re-estimating growth before 1820. Maddison-Project Working Paper WP-4

Broadberry S (2007) Recent developments in the theory of very long run growth: a historical appraisal (Warwick economic research papers, TWERPS, vol 2007, no 818). Working Paper. Department of Economics, University of Warwick, Coventry

Clark G (2007) A farewell to alms: a brief economic history of the world. Princeton University Press, Princeton

Cohen D (2009) La Prospérité du vice: Une introduction, inquiète. à l'économie. Albin Michel, Paris

Colletti L (1974) From Rousseau to Lenin: studies in ideology and society. Monthly Review, New York

[29] Ricardo and Keynes had one more in common: Pierro Sraffa. He was a friend and colleague of Keynes at the University of Cambridge, and also the master editor of Ricardo's Works, finally published in 1973.

[30] Keynes J. M., *Essays in Persuasion* (Keynes 1984, p. 179).

Diamond J (2005) Collapse. Penguin, London
Findlay R, O'Rourke K (2007) Power and plenty: trade, war, and the world economy in the second millennium. Princeton University Press, Princeton
Galliani F (1751) Della Moneta (On money)
Heilbroner R, Milberg W (2008) The making of economic society. Pearson Education, Upper Saddle River, NJ
Himmelfarb G (2008) The roads to modernity. Vintage Books, London
Hindes B, Hirst PQ (1979) Pre-capitalist modes of production. Routledge & Kegan Paul, London
Keynes JM (1984) Essays in persuasion. The Royal Economic Society/Cambridge University Press, Cambridge
Lefebvre G (1962) French revolution. Columbia University Press, New York
Malthus T (1985) An essay on the principle of population. Penguin, London
Mill JS (1981) On liberty. Penguin, London
Monroe AE (2006) Early economic thought: selected writings from Aristotle to Hume. Dover, New York
Smith A (1986) The wealth of nations. Penguin, London
Trevelyan GM (1946) English social history. Longmans, London
Tainter J (1988) The collapse of complex societies. Cambridge University Press, Cambridge
Voltaire (1979) Essay on the manners and spirit of nations. In: Redman BR (ed) The portable Voltaire. Penguin, London

Crises and Theories After the Industrial Revolution

8

Abstract

The optimism of Adam Smith did not exactly resonate with the reality of markets: persisting imbalances between supply and demand quickly led to the first capitalist crises. With falling profits, wages were further squeezed and soon a new movement was born demanding better conditions for the working class. Two formidable Germans interpreted the crises as a structural weakness of the capitalist system that would inevitably lead to its destruction. Building upon the Hegelian premises of historical phases, the newly founded Marxist theory advocated a new economic order that eventually would replace markets and ownership and uproot the profit mechanism for the benefit of the working class. The second half of the nineteenth century is marked by the second wave of industrialization, deep social divisions on how output is distributed and new monetary structures to achieve better coordination worldwide. New schools of economic thought in Germany and Austria dispute the prevalence of economic process and focus on the importance of history and the actions of individuals as the engine of growth. In the fast-rising US, economists search for how to stabilize wages and prices. Disapproving of the unproductive wealthy classes, Veblen supported the new strata of engineers and managers as the driving force towards more efficient production.

8.1 The First Crisis of Capitalism and the Emergence of Socialist Ideas

It was in 1825 England that the first, post Industrial Revolution crisis took place and quickly spread as a debt crisis in overseas states. Convertibility to gold had been suspended since 1797 as in the wake of Napoleonic Wars military expenses were financed by massive issuance of new currency, thus spreading the risk of panic should there be a rush to the banks. After emerging victorious from the Battle of Waterloo in 1815, England wanted to restore the convertibility law, and the central

N. Christodoulakis, *How Crises Shaped Economic Ideas and Policies*,
DOI 10.1007/978-3-319-16871-5_8

bank had accumulated enough reserves to make the system of convertibility credible again.

Monetary restoration proved a great success and was followed by a period of rapid economic growth. Gold reserves were abundant thanks to booming exports, and the British government increased public spending and investments in infrastructure. Lending from private banks also boomed and several of them were allowed to print their own currency, increasing monetary liquidity even more. Credit was also readily available to the newly independent nations of Latin America, thereby creating an international bond market in London. As the fear of a potential panic had eclipsed by then, the Bank of England further increased money supply, allowing the government to repay a big part of debt that was accumulated during the war. Meanwhile, share prices at the London Stock Exchange were soaring, with investors rushing to buy the stock of promising companies that were advertising generous returns in the future.

In 1825, however, investors were dismayed to discover that many stock companies were either a hoax or would never be able to deliver the high returns they were promising, and the market crashed. Dozens of private banks collapsed, causing a bank-run as depositors desperately tried to convert their money into gold before it was too late. The Bank of England took action as a lender of last resort by liquidating its gold reserves but it was overwhelmed by demand and would have collapsed too, had it not gained emergency access to the reserves of the Bank of France. In an ironic twist of history, a defeated nation—but since then adopting prudent accumulation policies—saved the rejoicing victors!

The Bank of England took measures immediately: it raised interest rates, put breaks on credit and forbid private lenders from issuing currency. Liquidity disappeared, forcing several Latin American debtor countries to default as cash was no longer forthcoming from London. Liquidity crisis was followed by massive business failures and the real economy entered a deep recession with soaring unemployment and a dramatic deterioration of working conditions. Thus it became painfully clear that the new era of capitalism didn't only concern new employment opportunities, ever-rising profits and unhindered development; the maladies of crisis, systemic failures and rampant poverty were also intrinsic, not foreseeable by the founding fathers and not preventable by the invisible hand.

The conditions in the labour market and the blatant exploitation of workers began to be dully scrutinized by economists, politicians and philosophers. A new movement demanding the improvement of work conditions, wage increases and universal suffrage[1] quickly emerged and gave birth, a few years later, to the Chartists, who in 1842 organized the first general strike on record. The Chartists were also at the forefront of the labour mobilizations that took place later in 1848.

A remarkable personality of the time was Robert Owen (1771–1858) who envisioned a new industrial society with less exploitation and a better life for

[1] That is universal manhood suffrage, as political rights for women were inconceivable at that time. It took more than one and a half century for societies to adapt.

workers. To implement his ideas, he established a model industrial area on his own property, where workers were provided with accommodation, education and even participated in managing the plant. Owen developed his principles in several books and lectures, contributing to the movement of "utopian socialism", initiated by Henri Saint-Simon.[2] He presented practical solutions to various issues, including the improvement of labour market functioning and legitimizing abortions so that poor families could avoid having too many children. His solutions tried to raise the responsibility of the workers, rather than treating them as an undignified crowd as pictured by Malthus. Owen devised rules for wage-setting in the public sector and was in favor of public investment to create more jobs, foretelling some Keynesian prescriptions that followed a century later.

The city of Manchester was the industrial heart of England having a large number of steel and textile factories, thus soon became the bulwark of the workers' movement. When the son of a German industrialist took over the management of a factory, he was shocked by the harsh working conditions. He approached the Chartists and in 1844 published his account in a book on "*The Conditions of the Working Class in England*". His name was Friedrich Engels (1820–1895), later becoming a close associate of Karl Marx (1818–1883) who was also studying the working conditions but from a far more radical perspective. Together the two would devise a brand new theory to interpret exploitation and economic crises, while also proposing an alternative model of organizing production and governing society.

8.2 The Emergence of Economic Thought in Germany

It was not until late-nineteenth century that Germany became a unified state, though even after this development, it still kept the structure of federal states, the *Ländern*. Contrary to England and, to some extent, France, Germany was not at that time focusing on mass industrialization and the role of central government was neglected. Karl Kautsky, a leading Marxist, aptly remarked that "[i]f 18th-century France was one hundred years behind England, Germany at that time was one hundred years behind France".[3]

Consequently, most German economists were concerned with the agricultural production and how to maintain the balance with demand in local markets. Though their findings were neither elementary nor ephemeral, the issues were suited to a community-based economy rather than to an emerging economic power. Karl Rau (1792–1870) developed the supply and demand diagram in its present form, while von Thunen presented a locational model of "spatial variation" for the use of agricultural land in which cultivations around the city are organized in concentric

[2] Owen, *A new View of Society* (1813).

[3] Kautsky, *The Materialist Conception of History* (1902).

circles.[4] Another economist of the time was Georg F. List (1789–1842) whose contributions were split on whether to view Germany as a federation of small states or a unified economy; characteristically, he was at the same time a proponent of free trade between German states and also in favor of tariffs on imports from abroad.

German economic thinking might not have been impressive, but history and philosophy were advancing fast and fabulously. The leading figure was Georg Hegel (1770–1831) who developed the "dialectic method" to account for the historical course of human societies. According to the Hegelian dialectic, human societies evolve through conflicting aspirations and actions. Each proposition is a 'thesis', the reaction to this is the 'antithesis', and the outcome of their interaction is the 'synthesis'. The scheme was quite radical for its time, as it presented society as being in a state which was *de facto* destined to evolve into a higher level, overcoming the obstacles that kept it stagnant. In his words, "[t]he mutations which history presents have long been characterized in general, as an advance to something better, more perfect", in contrast to the repetitive cycles taking place in Nature.[5]

Obviously, Hegel was responding to the backwardness of Germany relative to the other European powers. His theory inspired people to change existing conditions and legitimized attempts to overthrow authoritarian regimes in order to achieve social progress.

But as social change was becoming irresistible, two fundamental questions emerged: who are the agents of expressing the "antithesis" and who will take up the task of bringing about the much wanted "synthesis"? With regard to the first question, Hegel argued that an intellectual *avant-garde* should be formed, capable to shape and transmit the spirit of the times (*das zeitgeist*), typifying the vision and the value-system of the historical period. In ancient Athens, this role was played by army generals and philosophers, while Athens itself was the *zeitgeist* for the entire ancient world. In later eras, the avant-garde was in turn personified by knights, monarchies and enlightened despots. For the world in general, Hegel assumed that Germany was destined to be the engine of progress.

Even if the first question could be answered in such a manner, the thornier issue yet was the outcome of "synthesis". In addressing it, Hegel divided the progression of history into "historical stages". Each stage occurred when specific preconditions were met and its own agenda was established, before passing the torch to its successor in the unstoppable march of history. This concept gives Hegelian theory a teleological character, by which a "Messianic mission" that would allow the pursuit of the ultimate goal ("*telos*") is justified.

---

[4] Farmers grow vegetables in the fields closest to the city, while the more distant fields are used for more durable goods that last longer and can be safely transported over greater distances. For a description of von Thunen's model see Blaug, *The economics of Johann von Thunen* (1985).

[5] Hegel, *The Philosophy of History* (1956, Introduction, III, p. 54).

It is easy to see why these two tenets of Hegelian theory influenced a wide range of diverse movements and ideologies, from lofty Romanticism and idealistic naturalism to the doctrine of "Pan-Germanism"[6] as well as international Marxism.

8.3 Marxist Economic Theory

Hegel's theories inspired a new approach to economics, called "historicism", according to which institutions, social structures and behaviors are the evolutionary outcome of specific historical paths followed in the past. In this context, there are no natural, and hence universal, economic laws but every interpretation should take into account the stages of progress realized in the specific society. The same pattern is widely evident in Marxism as well.[7] Deeply influenced by Hegel's dialectic approach of historical progress, Karl Marx extended its scope and formulated a new version called "dialectic materialism". The innovation was that it is now the material conditions prevailing in each socioeconomic structure rather than abstract concepts governing the progression of history. No other economic theory has inspired more praise or caused outright condemnation as Marxism and this happened mainly for two reasons:

First of all, because it openly endorsed and herographed a specific social group—the working class—as the force destined not only to liberate itself from exploitation but also relieve society as a whole from alienation and enable it to reach its full potential, hitherto enslaved by the existing modes of production.

Secondly, because Marxism argued that economic crises are intrinsic in all modes of production. Crisis dynamics are at times manifested in a controllable manner demonstrated by the deprivation of workers, while at other times they become explosive, causing the collapse of the system in question, except communism, the system that will eventually replace them all.

8.3.1 Key Concepts in Marxism

Marxist theory was based on concepts of economic analysis that Marx either discovered and developed from older theories or devised himself and/or jointly with Friedrich Engels. Some key aspects are the following:

(a) *The distinction between work and labour power*: According to Marx, the former is the amount of labour actually put into production and is incorporated in the final product as "labour value"; labour power on the other hand, is the capacity of the worker to sustain and reproduce working labour, and is

[6] The latter dominated Germany during the nineteenth century, leading to its unification. Later it was also tried abroad, legitimizing the German occupation of other states in the course of two World Wars.

[7] In their work *Economies and Cultures: Foundations of economic anthropology* (2007), Wilk and Cligget consider Marxism as a product of German Romanticism. Their view is verified by a number of physiocratic positions they endorse, and by the manifest similarity between Marxist theory and the German School's historicism.

compensated by wages. The value of work is always higher than the value of labour power; the difference being what he called 'surplus value' and is appropriated by the capitalist as profit. Marx was greatly concerned about the "iron law of wages", a concept initially proposed by socialist and philosopher Ferdinand Lassalle (1825–1864). According to it, real wages always tend to decrease, in order to compensate for the downturn of capitalist profits facing the diminishing rate of return. In this context, Marx was the first to analyze the consequences of the "reserve army of unemployment" as the mechanism that is used to depress wages. His claim, however, that "the general tendency of capitalist production is not to raise but to sink the average standard of wages",[8] was disproved by the rise in workers' remuneration in the capitalist countries over the long term.

(b) *Organic composition of capital*: Marx made a distinction regarding the composition of capital, between the technical side and that in value terms. The latter is the ratio of the value of materials and fixed costs embodied in the production of a commodity to the costs of the labour power employed in producing it, i.e. the wage bill. The "technical" composition refers to the definite quantity of labour power required to produce a specific quantity of products.[9] This distinction would often resurface in later economic theories concerning the nature and value of capital, as for example in the Cambridge capital controversy which will be discussed in Chap. 13.

(c) *Tendency for the rate of profit to fall*: Since capital has the tendency of diminishing internal rates of return, capitalists are always striving to further surplus value, leading to an endless investment in fixed capital which inescapably results in falling rates of profit. To compensate for this loss, capitalists exert further pressure on wages and when exploitation exceeds boiling point, the system will finally collapse.

(d) *Markets and theory of money*: Contrary to Adam Smith, who rather overlooked the role of money, Marx acknowledged it as a type of "exchange commodity". The evolution and expansion of transactions extenuate the contrast between the "exchange value" and "use value" of commodities, and this trend upgrades the role of money.[10] Engels had also attempted to explain the exchange role money plays, defining it as a universal commodity through which all others could be exchanged. He went further to stress the special role and immense power of the financial sector in the economy, arguing that "when men invented money, they little suspected that they were creating a new social power, the one universal power to which the whole of society must bow".[11]

Marx was not, however, as insightful in understanding the role of the market. He thought of it as just a mechanism of transacting goods and disregarded its

---

[8] Marx, *Wage, Labour and Capital* (1847).

[9] Marx, *The Capital* (1954, Book A, XXV).

[10] Marx, *op. cit.* (Book A, I).

[11] Engels, *The Origin of the Family, Private Property and the State* (1884).

other functions on conveying information to consumers and producers alike. This neglect would prove fatal for the economic system that claimed to apply Marx's visions half a century later. By ostracizing the circulation of money and the functioning of markets, the system produced devastating consequences to the economy and the very survival of workers who were supposed to be the chosen few.

(e) *Economic fluctuations*: Marx used his assumption about the insatiable need of capital accumulation to develop a systematic and original theory of the business cycle. As accumulation leads to overproduction, excessive supply—combined with decreasing demand because of depressed wages—will cause prices to plummet. This is immediately followed by a dramatic fall of the rate of profit and business failures occur at a massive scale. These failures decrease the stock of capital employed in production, which now symmetrically leads to a rise in the rate of profits. A new cycle then begins, which lasts approximately 10 years before repeating itself all over again.

(f) *The primacy of economic structure*: According to Marx and Engels "the theories of the ruling class are in each era the dominant theories as well, i.e. the class which is the ruling material force of society is at the same time its principal intellectual force".[12] The implication is that there are no natural laws in economics, just mere class dynamics that shape the prevailing economic relations as the ruling class establishes an institutional framework to safeguard its interests. This idea gave birth to a new current of political and economic thought, which relegated all social phenomena—from religion and family to education and culture, to mention only the main ones—to mere "derivatives" of class struggle. As will be examined in Chap. 10, Marxist reductionism that "the economy reigns supreme" shares some common features with the formalism of neoclassical theory, according to which rational choice in terms of utility maximization can be used to explain the entire spectrum of social behavior and interactions.

Based on these concepts, Marxism developed a teleological theory predicting the ultimate collapse of capitalism and its replacement by the communist system of production. The main arguments are as follows:

8.3.2 The Collapse of Capitalism

Marxist theory was frequently adopted by the organized worker movement—to various degrees and not without internal divisions—both in demanding a better life and improved working conditions, and as the harbinger of a new society which would ultimately make their struggle moot by rooting out the causes of their plight. For the second goal to be worth pursuing, the first one should prove impossible to

[12] Marx and Engels, *The German Ideology* (1846).

attain on a permanent basis. In other words, whatever improvement workers can achieve in capitalism is temporary (if not illusive), so overthrowing capitalism was a choice inevitable as well as morally necessitated.

By extension, capitalism was seen as being in an intrinsic and unstoppable state of crisis, and here is where the first problems of Marxist crisis theory start to appear. For despite the insightful identification and account of economic crises, Marxist theory became self-constrained and dismissed any chance of renewal of the capitalist system of production. For this reason, Marxist theory was often convincing in explaining the root-causes of a crisis, but almost never capable in rightly assessing its long term consequences.

A central paradox in Marxist theory was generated by assuming a permanent inability of capitalism to adjust, in contrast with previous modes where transition was the norm: from slave ownership to feudalism, from feudalism to capitalism. However, capitalism had no other option but to collapse and make room for socialism. Thus the economic system that resulted in unprecedented levels of political freedom to the masses and had dramatically improved the material conditions of their lives (as Marx's own analysis confirmed) did not have, according to the same analysis, any potential for further evolution and transformation.

This was too often disproved by reality. Refuting Marxist predictions, the capitalist mode of production took many alternative forms; corporate structures growing from local enterprises to multinational giants; state intervention ranging from letting market activity uninhibited to setting-up of publicly owned industries; inequality varying from abject exploitation to generous welfare systems. In all its incarnations, capitalism proved remarkably resilient, not hesitating to provide opportunities even to things threatening the very heart of the system. Eventually, Marx's joke about the last capitalist who would sell the rope to be hanged by the masses was turned upside down: in capitalism, the last buyer would be more likely to sell the rope at a profit without anyone having time to use it.

Not even the root-causes of the crises were always correctly pointed out, as became evident when Marx and Engels tried to interpret a commercial downturn in England during the 1850s. They rushed to attribute it partly to the stock price speculation on railways and partly to the overproduction of agricultural products that led prices to plummet. In actual fact it was that the sovereign bond market had become too risky and surplus capital found an outlet in industrial production which grew steeply, causing a sharp fall in prices. This coincided with the abolition of corn tariffs, causing agricultural prices to fall as well. It is striking that Marx and Engels gave so much emphasis to overproduction that they overlooked the benefits of falling prices for the purchasing power of the population at large; a fact disproving the Marxist thesis that living standards of the masses constantly deteriorate to make room for keeping profits unscathed.

8.3.3 Structure of the Communist Economy

The economic theory developed by Marx and Engels was castigating capitalism in every possible manifestation but on the system that would come to replace it no specific outline was uttered. Several great economic thinkers were fortunate enough to witness the application of their theories during lifetime. For example, Adam Smith saw the rapid growth of the market economy, Ricardo witnessed the abolition of tariffs, Keynes saw his advice about state intervention heeded by policy makers, Friedman's theories were frequently adopted by monetary authorities, and the same goes for Mandel's theory on monetary unions. In this way, the founding fathers had not only the satisfaction to see their ideas being put in practice but also the opportunity to intervene and comment upon. The founders of Marxist economic theory, however, did not live long enough to oversee the implementation of their theories in any society.[13]

Their argument on the communist mode of production and the interim socialist system focused mainly on how workers would benefit, since the exploitation and the mechanisms oppressing their class would have come to an end. However, on the crucial point of how production and distribution should be organized under communism, original Marxist theory was vague and uninformative. The undisputed legacy they left was that capitalism would be ultimately replaced by communism; a system where the working class takes over production and operates under the principle "from each according to his ability to each according to his need". Communism presupposes the abolition of the state and its authority, but as this goal was not considered immediately feasible, an interim "socialist state" is necessary where individual property of the means of production is abolished and substituted by state property, thus the unsavory capitalistic profit is uprooted anyway. Some core principles are prescribed in the *Communist Manifesto*, fervently written in the midst of the social revolutions of 1848 in Europe.

Thus the question of how exactly to implement Marxism was left for self-appointed successors to provide an answer. The latter formed numerous schools of thought and proposed a great number of alternative practices. Internal conflicts were frequent, sometimes violent and deadly, as each one claimed to represent the original Marxist theory and authoritatively express the vision and prognostications of its founders. The most extensive and long-lasting economic system implemented in the name of Marxist theory was the model of centrally planned economy adopted during the twentieth century initially in the Soviet Union, and later in China and several East European countries.

Some argue that none of the analysis and evaluation of this system say anything about the original Marxist theory; they also claim that central economic planning and the uncontested rule of the communist party are alien to the theory of Marx and Engels. Nonetheless, the communist regimes are the only real life cases that can be used to evaluate communism empirically. All alternative approaches are

[13] Albeit a brief chance provided by the Paris *Commune*, in 1871.

hypothetical and imaginary, having no practical use for testing and evaluating the theory. A crucial aspect to test the validity of the Marxist approach is to check whether the root causes of the capitalist crises were eradicated by the new system that was established in the name of the founders. The central assumption is that they are caused by the insatiable thirst for capitalist profits, thus by controlling the accumulation of capital the diminishing rate of return will also be checked, and crises will no longer appear. Fifty years later, it was Stalin who implemented the new economic system in the name of Marxism and lost no time in proclaiming the end of the crises by promising that:

> [i]nstead of development of production with breaks in continuity from boom to crisis and from crisis to boom—unbroken expansion of production; instead of periodic breaks in technical development, accompanied by destruction of the productive forces of society—an unbroken process of perfecting production on the basis of higher techniques.[14]

History, however, did not validate such expectations and the system suffered from many crises before ultimately collapsing in 1989 as will be examined in more detail in Chap. 11.

8.4 Crises and Economic Unions

Following the crisis and revolutions of 1848, Europe entered a new phase of economic development with international trade growing rapidly and Industrial Revolution spreading to many countries. As trade balances were interacting with the accumulation of gold reserves according to the species flow model, rising economies in continental Europe quickly understood that they had to curtail dependency on US and British monetary policy and set up their own currency system. Finally, France took the initiative for the establishment, in 1866, of the Latin Monetary Union (LMU) consisting of ten European states, plus Venezuela who followed Spain. The currencies of member states were still sovereign, but all meant to be interchangeable at a stable exchange rate between gold and silver coinage.

Instead of cementing stability worldwide, the establishment of independent monetary systems without any sense of global coordination soon led to international crises. The LMU faced its first test when "Iron" Chancellor Bismarck unified Germany in 1870. As Constantine the Great before him, Bismarck decided to safeguard the credibility of the new German currency by pegging it to gold and abolishing the circulation of silver coins (the *taller*). The decision hit demand for silver and caused a steep drop in its price in international markets. A crisis mechanism was then set off as other nations immediately felt the consequences:

In the onset, it was the United States that in an attempt to prevent inflation by the circulation of cheaper silver coins abolished their convertibility to gold. As the

[14] Stalin, *Economic Problems of Socialism in the USSR* (1952).

price of the latter skyrocketed, this was tantamount to a drastic monetary contraction and deflation. Global and local demand fell, several US banks, whose reserves were mainly in silver, collapsed and silver mines shut down causing further unemployment and recession. England and other parts of the world were also hit, leading to a major international crisis described as the "Great Depression", until an even more severe crisis in the 1930s claimed this status.

The most remarkable developments, however, took place in the Latin Monetary Union. The Vatican took advantage of falling prices and minted more silver coins that they would still trade for golden ones of other states, thus making enormous profits. After the initial shock, the Union forbids the circulation of Vatican currency, bimetallism was abandoned and convertibility was based only on gold reserves.[15]

The economies in Europe were also hit, experiencing soaring unemployment and social upheavals. Governments thought that they could return to stability by bolstering their gold bullions and believed—as the old doctrine of mercantilism suggested—that achieving trade surpluses was the way to do it. Thus, many countries, including France and Germany, took a strong protectionist view by imposing tariffs on imports and subsidizing exports. Protectionism put the brakes on the internationalization of economies, but only temporarily. In several cases it motivated the accumulation of national industrial capital since domestic markets were still nascent and there was plenty of growth potential.[16]

Eventually the fact that three economic areas (the US, England and the LMU) were all operating under a Gold Standard meant that international monetary stability could be safeguarded and the ensuing flow of credit facilitated worldwide recovery. This led to the first wave of globalization, and economic theory began dealing with new problems such as monetary and fiscal coordination and international cooperation. Progress was so manifest that the period became known as the *Belle Époque*, to end only when World War I broke out. The Gold Standard was again reinstated in early 1920 with the establishment of a new and more unified—though shorter-lived—system that ended ignominiously and, in many ways, precipitated World War II.

8.5 New Schools of Economic Thought

By the late nineteenth century, economic thinking in Germany was crystallized to the Historical School; Austria nurtured the Austrian School of economics, while England and America were experiencing the "marginal revolution". It was during that time that economics became an independent field of knowledge, though its claim to scientific prowess was still in doubt.

[15] For an account see JCCB, *A History of Banking in all the Leading Nations* (1896, Part VIII).

[16] Almost a century later, similar theories would prevail in Latin America with disastrous consequences, as explained in Chap. 13.

8.5.1 History Matters

The German Historical School influenced a whole generation of thinkers believing that the best source of knowledge about an economy or a society is its own history, whereas mathematical modeling and comparisons were of little value. Consequently, economic theory must be "deductive", solely based on specific observations, and then move on gradually to more general statements. A famous representative of the School was sociologist Max Weber who in his *Protestant Ethics and the Spirit of Capitalism*, attributes the economic supremacy of the United States and Northern Europe to the intrinsic values of Protestantism that differ from the more lax norms prevailing in other countries.

8.5.2 Freedom Matters

The Austrian School is essentially a version of the historical school, arguing that economic phenomena should be studied by analyzing the actions of individuals, rather than of collectivities. Like the German School, it was influenced by Hegelian historicism as well as by the trends in sociology and psychoanalysis that were flourishing in Vienna at that time. The most influential representative of the Austrian School was Carl Menger who dismissed quantitative economic analysis altogether, arguing that the behaviors of different societies are not comparable. Menger focused on the theory of money and capital, and became a staunch opponent of Marxist labour theory.

Another distinguished member of the School was Ludwig von Mises who took exception to both positivism which attempts to interpret specific phenomena based on general laws, and to materialism which robbed individuals of any meaningful role in shaping the course of history. He was an uncompromising proponent of free markets and he considered as one of the most celebrated epochs of human history "the period between the Napoleonic Wars and the first World War [because] the social ideal ... was free trade in a peaceful world of free nations".[17]

Later, Friedrich Hayek (NLE, 1974) would develop the anti-state theory of the Austrian School, arguing that the market is the only mechanism able to recognize the preferences of individuals, express this information through prices and motivate production and supply. Based on this premise, he fervently fought against central economic planning arguing that it is impossible to determine prices in a socialist society due to the absence of a market mechanism. Any attempt to replace markets with the decisions made by commissars, deprives individuals the freedom of choice and subjugates them to a central authority.

[17] von Mises, *Liberalism: The Classical Tradition* (2005, p. xiii).

8.5.3 Economics Matter, in the United States

After the Civil War (1861–1865), the United States set upon the path of becoming a unified economy. The construction of extensive railway networks facilitated transportation, allowed integration of commercial activities and created unprecedented economies of scale. Ultimately, economic activity became concentrated with a few corporate behemoths dominating and creating the first cartels or business trusts. The need to form a big domestic market, able to consume the products of the growing industrial sector, resulted in a conflict between landowners and industrialists. Compared to a similar conflict in England half a century earlier, the interests of the two camps were now pointing to opposite directions. In the American case, it was farmers who were against tariffs on industrial products as the additional cost was passed on to them, while industrialists were in favor of protectionism since it served the growth of their own businesses.

As a result of these developments, new economic theories emerged focusing on the large size of enterprises and the problems caused by vested interests. Five important economists emerged at that time in the United States:

Simon Newcomb (1835–1909) was an astronomer and part-time economist, like Jevons. He developed a theory on how excessive price volatility can hurt the economy. To avoid sharp fluctuations in the purchasing power of the labor force, he suggested the introduction of an artificial currency based on the prices of a basket of goods that should be used for determining wage levels. Essentially, this is what we would call today a system of indexing wages to inflation, to ensure that real income levels always correspond to a specific amount of purchasing power.

John Clark (1847–1938), on the other hand, developed a theory arguing that value is a social phenomenon determined collectively and not by individual choice. Since he lived in a time when industry barons used ruthless practices to set foot in and control new markets, he distinguished between two types of competition: the "constructive" type is benign as it leads to lower prices and benefits consumers; "destructive" competition on the other hand destroys businesses and harms consumer interests. He was in favor of mandatory arbitration and worker participation in corporate profits, and argued that payments to the factors of production (land, capital, labor) should be determined by their marginal productivity.[18] Clark, however, dismissed the importance of business cycles, which is why he failed to grasp the causes of the 1893 crisis in the US. According to Clark, the recession was caused primarily by high prices hurting disposable income and domestic demand, while the fact that Europe was also in recession meant that the US could not export their way out of the crisis.

Thorstein Veblen (1857–1929) was interested in crisis mechanisms and the role that different social classes played in the economy. He is considered as the founder of institutional economics and also as an implacable denouncer of the rich. In his

[18] Clark enjoys a mixed reputation in the history of economics; opinions regard him from an apologist of capitalism to a pioneer of labour movement. For a presentation see Leonard (2003).

famous work *The Theory of the Leisured Class*, he echoed the Aristotelian dichotomy and considered that the economy is based on two different processes; one is the acquisition of resources that leads to the 'business' sector of unproductive activity, and the other is production by the 'industry' sector. In this context, he argued that engineers and managers both serve and practice capitalism, so they should be in charge of the transition to a new system.

The leisure class is defined as the propertied but non-industrial parts of society that thrive upon exploitation without providing any serviceability to society.[19] Therefore their consumption can lead an economy to recession, predicating related theories developed later by Kalecki. Veblen studied various phenomena related to the leisure class, such as speculation and market manipulation and suggested the establishment of strict regulatory institutions to deal with them.

Irving Fisher (1867–1947) worked on mathematical economic models and how marginalism is implemented in economic theory. He formulated the quantitative theory of money with great accuracy, defined real interest rate as the yield an investor receives allowing for inflation, and investigated the relation between capital and interest rates. His theory greatly influenced modern monetary theory and Milton Friedman in particular; for a recent account see Bordo and Rockoff (2011). With his analysis on the deflation effect on debt, he also influenced J. M. Keynes as examined in Sect. 10.2.2. Fisher was a supporter of economic reforms that serve price stability and dealt with the economic impact of wider social issues, including health, world peace and alcoholism.

Finally, John Commons (1862–1945) was a radical economist mainly following the German Historical School. He focused on the issues of taxation and social security, and stressed the importance of rights and institutions for economic activity.[20] He examined in detail the role that investment in public infrastructure can play in bolstering the economy and creating jobs. Commons was therefore a major influence on the New Deal policy that was presented in 1933 by President Roosevelt to drive the US economy out of depression.

References

Bordo M, Rockoff H (2011) The influence of Irving Fisher on Milton Friedman's monetary economics. NBER Working Paper Series, No. 17267

Blaug M (1985) The economics of Johann von Thunen. Res Hist Econ Thought Methodol 3:1–25

Chavance B (2012) John Commons's organizational theory of institutions: a discussion. J Inst Econ 8(1):27–47. doi:10.1017/S1744137411000336

Engels F (1884) The origin of family, private property and the state. Hottingen-Zurich. Available at https://www.marxists.org/archive/marx/works/1884/origin-family/

Hegel G (1956) The philosophy of history. Dover, New York

[19] Veblen, *The Theory of the Leisure Class* (1994, p. 129).

[20] For a recent appraisal see Chavance (2012).

JCCB (Editor of the Journal of Commerce and Commercial Bulletin) (1896) A history of banking in all the leading nations, vol 3 (France, Italy, Spain, Portugal, Canada). Available at http://oll.libertyfund.org/titles/2239

Kautsky K (1902) The materialist conception of history. Available at https://www.marxists.org/archive/kautsky/1902/08/aims-limitations.htm

Leonard T (2003) A certain rude honesty: John Bates Clark as a pioneering neoclassical economist. Hist Polit Econ 35(3):521–558

Marx K (1954) The capital (Das Kapital). Lawrence & Wishart, London

Marx K (1847) Wage, labour and capital. Available at https://www.marxists.org/archive/marx/works/1847/wage-labour/

Marx K, Engels F (1846) The German ideology. Available at https://www.marxists.org/archive/marx/works/1845/german-ideology/

von Mises L (2005) Liberalism: the classical tradition. Liberty Fund, Indianapolis

Owen R (1813) A new view of society. Cadell and Davies, London

Stalin J (1952) Economic problems of socialism in the USSR. International Publishers, New York

Veblen T (1994) The theory of the leisure class. Dover, New York

Wilk R, Cligget L (2007) Economies and cultures: foundations of economic anthropology. Westview Perseus, Boulder

From Accumulation to Distribution

9

Abstract

The founding fathers of economics were too concerned with the process accumulation and production to pay attention on whether the product is fairly distributed to the various social classes. Utilitarianism appears in the nineteenth century as a new theory that attempts to resolve the issue, but agreement was not easy. Should society pursue a collective prosperity as advised by Bentham or each person's utility is inalienable and cannot deteriorate to improve a third party as ruled by the Pareto criterion? In the course of history, utilitarianism served as the basis of the welfare state but also provided arguments for those pursuing sectarian policies in the name of society. In the meanwhile, a middle class is emerging and this makes the polarization between capitalists and workers to look less threatening than in the early years. Capitalism proved to be a lot more resilient than the Marxist prophesies foresaw, but not as much in equilibrium as envisaged by its own theorists. A succession of peaks and crises is becoming a common characteristic of the developed economies, and a constant challenge to economic theories.

9.1 The Role of the State in Classical Economics

The rise of the nation-state after the peace treaties of Westphalia in the seventeenth century had profound implications on the allocation of power and the emergence of new institutions to replace the *ancien regime*. The most characteristic changes included:

(a) The formation of homogenous societies, so as to thwart domestic conflicts between nationalities.
(b) The weakening of Church power, in order to control domestic religious antagonism.

N. Christodoulakis, *How Crises Shaped Economic Ideas and Policies*,
DOI 10.1007/978-3-319-16871-5_9

(c) The establishment of state bureaucracies, to take over functions which by then were exercised by non-state agencies in the name of the sovereign.
(d) The separation of powers and the establishment of constitutional guarantees for constraining or outright abolishing the monarchies.
(e) The gradual establishment of universal manhood suffrage.

Reflecting the new realities of statehood along national lines, a new theoretical paradigm emerged in economics. This transition was neither immediate nor similar across states since, even in late eighteenth century, there were structures in Europe far different from a uniform nation-state model.[1]

Broadly speaking, however, the scale of economic activity since then coincided with state borders, and economic thought gradually emerged as a theory of national economies as it still appears in today's textbooks. Oddly enough, classical economic theorists had not fully anticipated how a centralized, national state could affect the functioning of the economic system; for example by redistributing income through welfare systems or by suppressing wage demands the state could reallocate wealth and influence social dynamics. It was not until the late nineteenth century that economic theories started to deal with the economic role of the state and appreciate the resilience and importance of its functions.

Despite being the first of the classical theorists, Adam Smith had a lot more to say about the economic role of the state as compared to later thinkers. According to Smith, the government should act as a guarantor of institutions, a mediator in disputes and a provider of public goods; its obligation was to let the justice system functioning, ensure political stability and provide domestic security. In the economy, the state was advised to adopt progressive taxation and impose tariffs on the use of public goods, though the government should be under no obligation to secure employment of its citizens.

Unlike Adam Smith, Malthus neglected the role a state can play in dealing with food and overpopulation crises, and argued that government policies may have adverse side-effects by helping the weakest members of society to survive, thus attenuating the problem. He even argued against government intervention in cases of natural disasters, since the latter was considered as a nature's mechanism to deal with its overexploitation by mankind.

Ricardo was also ambivalent on the role of the state and limited his analysis to the consequences of taxation. Regarding public spending, Ricardo thought it was ineffective in improving the wellbeing of society, since households—anticipating that the government will impose higher taxes in the future to service its debt—respond by increasing their savings. This proposition became known as the

[1] These included German agricultural communities, autonomous Italian republics and regions, the Swiss confederations, Church lands and principalities, numerous kingdoms, and still four empires in place, (Russian, Austrian, Prussian and Ottoman). Several European nations were still under the Ottoman rule.

"Ricardian equivalence" and in the 1980s became the rallying cry of neoliberal economists who dismissed government fiscal intervention all together.[2]

At the other end of the spectrum, Marx saw the state as a mechanism that is defending and perpetuating the capitalist system of production. It was Engels, however, who culminated this instrumental approach by arguing that the state had traditionally been an agent serving the interests of the ruling social class at the expense of all others. Ignoring transformation stages, even as a result of the change of the balance of power in society that themselves had previously described, Marx and Engels were adamant that the dissolution of the state is a key condition for the overturn of capitalism.

9.2 From Production to Distribution Theories

The role of the state was perceived in more constructive terms in late nineteenth century when economic theory tried to understand the process of distribution and how it can be used to benefit society at large. The problem of social disparities came to the fore in great part due to the advancement of Darwin's theory on the evolution of species which subsequently influenced economic thought and legitimized cut-throat competition and uninhibited self-interest, regardless of how the weaker members of society would be hurt. Adam Smith's invisible hand, which was supposed to bring balance between various and diverging individuals, was no longer a sufficient explanation for a society witnessing the emergence of powerful business trusts and the abject exploitation of workers. Solutions that would improve the life of the poor were urgently needed, though the question on "who decides what is good for society" remained an open challenge.

9.2.1 Utilitarianism

Any "common good" theory would have to begin by dismissing Locke's thesis that individuals should always enjoy absolute freedom.[3] The first to attack it was David Hume (1711–1776) who in his famous treatise on *The Original Contract* dismissed the concept of natural rights as superfluous and meaningless in a polity. Jeremy Bentham (1738–1832) attempted to identify this elusive "common good" by invoking the Smithian premise that all individuals have the same right to pursue their

[2] In his proof of the Ricardian equivalence, Barro (1974) assumes that all households are exactly the same. But the theory is in trouble if we assume that the state intervenes to support a poor household that will pay minimal taxes in the future, and taxes a rich household with minimal impact on its consumption.

[3] According to Kitromilides (*Political Thinkers of the Modern Era: Biographies and Interpretations*, 1998, p. 141), "Bentham dismisses the concepts of natural law, natural rights and social contract—in other words the classical principles of liberalism established by Locke—as unrealistic and redundant".

interests. Taking one step further, Bentham formulated a theory of utilitarianism by defining the collective interest of society as the sum of interests of its individual members. In doing so he dismissed Aristotle's criterion of morality in evaluating collective goals, as well as of Kant's categorical imperative, and endorsed the dictum that the end justifies the means. This positivist concept of common good was glorified by French philosopher August Comte (1798–1857) who categorically argued that "*...there are no 'rights', men enjoy only duties. Only society has rights against man. Man's only right is to do his duty*".[4]

The next issue was how to measure utility in order to evaluate alternatives and choose between. The advances in mathematics played a crucial role here, particularly the theory of differential calculus which provided economists with a tool to measure the marginal fluctuations of an economic value (be it cost, profit, utility, etc.). This development led to the establishment of a new approach, called "*equi-marginal*" school or simply 'Marginalism'. W. S. Jevons (1835–1882) extended the analysis by introducing the concept of marginal utility and finding a mathematical solution to maximize it (or equivalently minimize displeasure). With very few exceptions, such as the German Historical School and the doctrines of Austrian economists, there was hardly any economic approach that did not—to a lesser or greater extent—use this instrument to its analysis.

Bentham's vision was therefore equipped with a new method capable to define, account for, measure and compare alternative ways to serve the common good, the impact of which was expressed by the social welfare function. This new way to calculate collective utility gave birth to the concept of "social engineering", which immediately attracted all those who fought to improve living conditions in the mass societies formed after the Industrial Revolution. The establishment of shelters for the poor in England, the introduction of mandatory social insurance and basic education in Germany, and a multitude of welfare policies were inspired by the goal of serving the common good.

Yet in his enthusiasm, Bentham failed to identify the weak spot in his theory which was later exploited by his adversaries: if social welfare is the sum of individual interests, it seems legitimate to make some people worse off if, by doing so, overall conditions can improve. A philosophical debate concerning individual freedom and utilitarian coercion sparked off.[5]

[4] Comte, *A General View of Positivism* (1971).

[5] Brutal manifestations of this rationale are the forced collectivization in the 1930s Soviet Union and the mass killings of citizens during Mao's so-called "cultural revolution" in 1966. These policies were considered a "necessary evil" to foster the conditions that would bring much greater prosperity to those survived.

9.2.2 Mill's Escape Route: Individual Freedom and the Common Good

To resolve the impasse, John Stewart Mill (1806–1873) introduced a different approach to utilitarianism. Influenced by the prevailing political liberalism of his time, he believed that individual freedom is an equally important and inalienable moral value. He therefore endorsed Locke's premise regarding the prominence of individual liberty and tried to reconcile it with Bentham's notion of common good.

According to Mill, men do need to develop their own traits and preferences to avoid uniformity but individualism should always be contained within the limits posed by the liberties and interests of others. To mitigate a potential conflict between individual freedom and the collective good, Mill turns to the government: the state should be ready to intervene and keep control of strategic corporations, so that it is able to satisfy the needs of citizens and avoid their subjugation to the demands of private profit.

In Mill's mindset, an individual remains the uncontested owner of inalienable economic capital (both human and physical), and may concur to the redistribution of wealth but not forcefully. Communism is dismissed by Mill specifically because it abolishes any sense of individual initiative which is seen as the driving force of progress in Europe. He fears that if the state is the sole owner and dispenser of corporations and services, society will suffer a loss of freedom and choice.[6] Because of these ideas, Mill is considered the founder of *democratic socialism* which takes a distance from both communism and economic liberalism, as individual freedom is precluded by the former and the need of government intervention by the latter.

9.2.3 Pareto's Criterion: Dismissing Asymmetrical Change

Vilfredo Pareto (1848–1923) dismissed the very foundations of utilitarianism, not even accepting a comparison of two alternative states of collective utility. It only takes a single individual to become worse off by a proposed reallocation, to dismiss any change of the initial state regardless of the collective outcome. According to his famous criterion, redistribution of endowments is acceptable only if no one is made worse off and at least one is made better off, otherwise an attempt to serve the common good could turn against a disproving minority. Pareto's criterion was philosophically compatible both with Voltaire's principle that one person's freedom ends where another's begins, and with the concept of inviolable individual rights introduced by Locke.

Though elegant in theory, the application of Pareto's criterion proved controversial in practice. First of all, it is possible that even a redistribution policy improving the position of all the members of a group, may run into problems if it benefits some

[6] Mill, *On Liberty* (1981, p. 182).

more than others. No one's utility is worse off but inequality increases, hence many individuals may react to such a change either because of envy or in the expectation that more inequality is detrimental to other functions of the group. If a policy benefits someone more than it harms another, a second question arises whether could be a "compensation mechanism", redressing the losers and ultimately turning them into gainers as well.

9.2.4 The Birth of Welfare Economics

Inquiries into this issue led to new theories, giving birth to the so-called "welfare economics" that would exert great influence on the evolution and evaluation of economics, especially during the second half of the twentieth century. The combination of Benthamite utilitarianism with the Paretian principle meant that collective utility can become mandatory only if accompanied by a compensation mechanism addressing the grievances of individuals who would otherwise become worse-off. A well-known illustration of this principle is the construction of a new road which serves the interests of the community as a whole, but will harm the private property of certain individuals. The government compensates the owners of the expropriated property and recovers the money by the increased tax revenue brought about by the new activity. Alternatively, compensation can be covered by obliging the community that benefits from the new road to pay a usage fee transferable to the ex-proprietors.

Often, however, this compensation mechanism fails as, for example, in the following cases:

(i) When the affected parties' utility is negatively related to the improvement gained by others (e.g. due to envy) and compensation is not sufficient to cover both the marginal loss and the envy factor.
(ii) When affected parties are politically vocal, well organized and suffer an immediate loss, while those benefiting are not a coherent group due either to their large number or the fact that the benefit is long-term and not easily identified. A coalition of the affected parties can impose very steep terms of compensation, while those who stand to gain from the change cannot negotiate efficiently.

A classic illustration of this problem is the reform in pension systems, in order to make them less costly and ensure their viability. Pensioners living now and suffering an income loss will vote in the upcoming elections, while those who stand to gain have not yet been born. The conundrum is resolved only if the government acts as the voice of future generations. To identify what constitutes a "just compensation", a widely accepted concept was introduced by Abram Bergson in 1938 and further developed by Paul Samuelson. The Bergson—Samuelson social welfare function may take three forms[7]:

(a) The utilitarian or 'Benthamite' form where the welfare of society is the un-weighted aggregate of individual utilities.[8]
(b) The 'weighted' welfare model that gives different weights to the utility of each group member. This treatment may reflect social priorities, the balance of power between social classes, considerations of inter-generational solidarity, etc. For example a government may give more weight to the preferences and interests of its voters, or a political party may focus on improving the position of the weaker members of society.
(c) The 'one-dimensional cause' according to which suffices to improve the state of a single social group. This is an extreme version of the previous model based on the belief that the other members of society automatically—though indirectly—benefit from the policy as well. In Pareto's terminology, the goal is to increase general utility in absolute terms by making some members better off, while no member is made worse off.

9.3 The Extremes of Utilitarianism

In this context, when the goal of society is to improve the status of the "poor" or the "rich" at the expense of the other, we can identify four extreme cases and correlate them with specific regimes and policies ranging from the praiseworthy to the repulsive. One may at first wonder how it is possible for a utilitarian to ever be repulsive: alas, history has frequently shown that many disasters and crises were caused by self-proclaimed saviors vouching to work for the common good. The following cases are discussed:

9.3.1 Maximizing the Utility of the Have-Nots: "Support the Poor"

The most well-known approach is the concept of egalitarianism advanced by John Rawls (1921–2002).[9] According to this the general social wellbeing is enhanced by improving the status of its weaker members. An application of this principle was the positive discrimination policy adopted in the US by the Johnson administration in late 1960s. The goal of the policy, also known as "affirmative action", was to increase black American enrollment in universities so as to achieve a higher integration in American society. The special treatment of African-American candidates was meant to compensate for their subpar performance in university

[7] For a concise account see Johansson, *An Introduction to Modern Welfare Economics* (1991).

[8] Perhaps a more accurate term would be "egalitarian utility function".

[9] In his seminal book, *A Theory of Justice* (1999).

entrance examinations, which was attributed mainly to the deprivation of their family and the inferior social conditions. The policy angered conservative groups, who argued that in this way African Americans "stole" university positions from white candidates who had performed better in exams. As this policy was blatantly violating the Pareto criterion, so it was hurting social welfare.[10]

Critics also argued that the policy was encouraging minimal effort, ultimately hurting both the beneficiaries who had inferior qualifications and were not required to train more, and society that was thus receiving lower quality services. The counter-argument is that society will benefit in the longer term from the positive effects of African-American university attendance and by motivating many members of this community to receive better education and follow the example of successful peers, instead of fatalistically accepting a life in poverty without such options. Society may also stand to substantially benefit from the achievement and representation of more groups in public life, which would perhaps overshadow the potential downgrading of service quality.

9.3.2 Maximizing the Utility of the Elites: "Support the Rich"

The opposite approach, by maximizing the utility of only the most privileged members of society, has also been advanced in various eras. Especially after economic crises, it is often argued that providing support to the powerful elite is necessary in order to achieve the rapid re-accumulation of wealth, even if this means that the poor temporarily have to suffer more. From Middle Age feudalists to the American robber barons in the nineteenth century,[11] this approach has been conveniently welcome by the rich and powerful.

In the twentieth century a debate sparked among economists on the relationship between inequality and growth: whether it is positive or negative and about the direction of causality, as discussed in Sect. 13.1.4.

9.3.3 Minimizing the Utility of the Elites: "Suppress the Rich"

Under this approach, social progress is linked to the decreased affluence of the elites, even if this does not necessarily imply that the weaker part becomes better-off. This trend also prevails mainly after economic catastrophes when governments, incapable of dealing with the ramifications of the crisis in a constructive way, resort

[10] This objection was frequently refuted in court. As part of the affirmative action policy, an African American lawyer was appointed at the Texas DA office over a white lawyer who had succeeded in the entry exams. The white lawyer went to court, but the judge explained that a more representative justice corps enhances the social sense of justice and therefore promotes the social welfare. See Sandel, *Justice: What's the Right Thing to Do?* (2009).

[11] The so-called "robber barons" accumulated vast fortunes during the nineteenth century and wiped out small businesses.

to punishment tactics against the "haves" in the hope of satisfying the vengefulness of "have-nots" and make them forget their own suffering. In the course of history there have been quite a few manifestations, frequently supported by religious fanaticism. The most famous agitator was perhaps Girolamo Savonarola in late fifteenth century when Florence fell in decline, following the overthrow of the Medici and the occupation by France. The friar argued that God would be pleased, not if the wealth is redistributed to the poor as Rawls might have suggested today, but just by destroying the riches so that vanity disappears.

During the twentieth century, the argument frequently inspired class-conflicts and forced expropriation practices.

9.3.4 Minimizing the Utility of Have-Nots: "Suppress the Poor"

The opposite approach is to minimize the utility of the lower classes, surprising as this might sound to modern ears. Through the ages, it has provided theoretical justification to many social patterns. In antiquity, cynic philosophers argued that people can be "liberated" from material compulsions and achieve moral purity only through limiting their needs to a bare minimum. In the Christian tradition, asceticism taught that salvation comes with abstinence from material pleasures. During the thirteenth century, a monastic order in Europe (the *Flagellates*) preached that deprivation and self-humiliation are necessary conditions for redemption. Franciscan friars practiced absolute poverty and lived by alms alone, following the example of their founder, Francis of Assisi.[12] Needless to say that the ruling classes never felt compelled to follow such advice, while its endorsement was a convenient excuse for the impoverishment of their subjects.

Pro-slavery theories were by construction dismissive of the concept of individual utility. Legalistic arguments were used to justify enslavement either by birth (natural law), inherent inferiority (racial law), as a result of punishment (penal law) or captivity (law of war). In the absence of any such pretext, they were enslaved by brute force. Even after the official abolishment of slavery, theories justifying the subjugation and enslavement of others circulated until the mid-twentieth century, serving the interests of various self-proclaimed superior groups.

In another manifestation of such beliefs, theories about "the undeserving poor" were spread around Europe in late Middle Ages and survived through the modern era. Perhaps surprisingly, the proponents were not the old lords in a last ditch to deter the upcoming masses, but those who stood against feudal authority, including religious reformers. In the Edict of Worms where the key concepts of Protestantism were laid in 1521, Martin Luther famously proclaimed that salvation comes through faith alone (*sola fide*) without requiring alms giving or other contributions. The

---

[12] For a description of the foundation of the movement, see Nicholas, *The Evolution of the Medieval World* (1992, Part III, Chap. 12, Spiritual life in Middle Ages).

reason was entirely economic, since mandatory almsgiving imposed by the Catholic Church frustrated the rising classes of burghers; they preferred to finance the initial accumulation of capital and expand business and fortunes as quickly as possible. The fact that proceeds were collected and managed by the Holy See was only to make their rage more fervent. Charity was seen as an additional tax and, to avoid it, they doubted its expediency by openly asking if it was worth trying to help the poor instead of abandoning them to their fate.

The deliberate attempt to deteriorate the status of the poor was expressed as an economic theory by Malthus, who argued that purging the weak in order to decrease population and ease the pressure on food production turns in favor of society. Following the same logic, Malthus thought that wars and epidemics are processes serving the above need, and argued that the poor should be allowed to marry only in old age when is no longer able to reproduce.

The theory of the undeserving poor can be projected to entire countries. Karl Popper, for example, despite being a fervent advocate of liberty, was not equally unequivocal in his views about the national liberation of Third World countries. Commenting on the abject poverty of many nations, Popper embraced Hegelian historicism to argue that famines are the result of structural deficiencies in societies, and went on accusing the West of "prematurely liberating those nations".[13] Obviously he would rather see them remaining colonies until maturing enough, though the problem who would assess the degree of maturity was left unanswered.

Even more recently, the theory of the "undeserving poor" resurfaced in public policy debates as, for example, in the attack against the welfare state that was campaigned in the US during the 1980s.[14] In the same line, the republican presidential candidate in 2012 was reported as saying that he "did not care for the poor". Grilled by his opponents, he then clarified that he meant he was "not concerned with the poor", implying that the welfare system in the US is so effective that one does not need to worry about the problems of its claimants.[15]

9.4 The Classical Economists and the Social Classes

All major classical economists (Smith, Ricardo, Malthus), as well as Marx, focused their theories on the factors of production: how they are accumulated, in which way they can be optimally combined, and how they are remunerated. They were also concerned with whether and when these factors are likely to be exhausted, causing economic crises. Their analyses provided, explicitly or implicitly, an ethical and

[13] In an interview published in Der Spiegel magazine (13, April 1992). The excerpts are quoted in Popper's book *All Life is Problem Solving* (1999).

[14] For a description see Katz, *The Undeserving Poor: America's Enduring Confrontation with Poverty* (2013).

[15] A pertinent debate can be found in *The Economist* (24/1 and 1/2/2012).

ideological armory to the social classes which had either the ownership of, or the claim on, the factors of production.

To begin with, Adam Smith praised producers and specialized labor as the key to progress, while criticizing luxurious consumption and idleness of the wealthy as their extravagance deprived the economy of productive resources. Ricardo favored the industrialists and—indirectly—industrial workers against the landlords, whose rents claimed a disproportionate share of economic output. Malthus was weary of overpopulation and, to help society to survive within the production constraints, he supported the landlords as the owners of this scarce factor. Marx was the most outspoken of all, arguing that all production factors were in the past created through labour and thus belonged to the present and future working class.

All four approaches—despite methodological differences and social preferences—dealt with society at large and how it can prosper (Smith), survive (Malthus), operate more efficiently (Ricardo) or eradicate the sources impoverishing the workers (Marx). To accomplish such aims, society should follow a specific model of accumulation and deployment of factors of production according to the respective prescriptions. None of them, however, paid explicit attention to the role and behavior of the individual. Even in the work of Adam Smith, who is considered the ideological father of individual choice, labor specialization, production and consumption, are seen as the results of the prevailing circumstances, not the sole outcome of individuals exercising some kind of natural right.

Malthus is even more dismissive of the importance of individuals, seen as mere numbers that have to conform to whatever population limit is considered viable at a given time. Individuality is ignored (and frequently abhorred) by the theories of Marx, and this does not pertain only to the members of *bourgeoisie*. Not even workers are viewed as free competent individual agents (as required by Locke) or at least as self-interested units (as described by Hobbes). For Marx, the individual has economic value and acquires historical and class consciousness *only* as a member of class. Exposed to the relations universally imposed by other classes, each person has the historical mission to overthrow exploitative relations and establish a new system that will prevent them from re-appearing.[16]

Downplaying the importance of individual choice in the economy becomes even more paradoxical given that along the evolution of classical economics there was a parallel outburst of political movements in both Europe and the US demanding more rights and individual freedoms. Classical theories seemed to be sterilized from the intellectual influence of the Enlightenment, the defeat of absolutism, the rise of political liberalism and the waves of humanistic traits that were sweeping philosophy and culture. An important consequence was that the gradual rise of intermediate social classes went largely unnoticed by economic theories.

[16] According to Olson, this was the main disproof of Marxist theory, since the individual cost of undertaking such action is infinitely greater than the potential collective benefit. See Olson, *The Logic of Collective Action: Public Goods and the Theory of Groups* (1965, p. 104).

9.5 Theories of the Middle Class

One of the most remarkable socioeconomic developments of late nineteenth and—mainly—the twentieth century was the expansion and empowerment of middle social classes. This development had major consequences for both the supply and the demand side of the economy. During the initial stage of Industrial Revolution, society rallied around two camps: the upper bourgeois class (businessmen and merchants) that was growing rapidly thanks to the accumulation of capital and profits, and the much larger social group of industrial laborers. Workers performed their tasks in a more or less uniform way and earned more or less the same wages. There was no "middle class" worth speaking of, while the old nobility had lost its political power and was by all means trying to become accepted—economically and socially—by the new status quo.

9.5.1 Marxist Theories on the Middle Class

Marx developed his theories by focusing on the fundamental conflict between labor and capital in a highly polarized social environment. Later Marxist theorists would admit to the existence of divisions—and periodical conflicts depending on the business cycle—within the capitalist class, and also between the high and low middle classes. But this admission did not mean that Marxism had come to recognize some special—let alone critical—role to be played by the *petite bourgeoisie*, as the intermediate stratification was being scornfully referred to. The latter's existence mattered only insofar as there was a possibility of becoming a political ally to the genuine working class, in its inexorable path to assume power. At least for the early Marxists, the world was characterized by homogeneity and uniformity, both in terms of the production process and with regard to the social consciousness and political affiliations of its members.

The dynamics of industrialization, however, was not just about quantitative leaps in productivity but also exercised a constant pressure for innovation. Technical advances became a crucial condition for the expansion of industry, while the complexity of new methods meant that there was also a growing need for managerial competence and oversight. The monotonous uniformity of factory work was only part of the picture and waning over time; businesses increasingly needed able executives, while new methods allowed the measurement of value-creation and the contribution of each sector of the economy.

While Marx and Engels believed that this class was an intermediate stage for those being demoted from the ranks of the bourgeoisie, the middle stratum was in fact becoming a potential stage of promotion for those from below. The bipolar class analysis of Marxist theory was facing additional difficulties by the economic, institutional and technological developments, since actual world was not as clear-cut as assumed in theory. New social relations were emerging all the time, shifting the focus from the original conflict between labor and capital to other antagonisms, including those between moderate and massive accumulation, national and foreign ownership, specialized and unskilled labor, urban and rural activity, and so on. Gradually, the middle class acquired more rights, owned houses and established small businesses, while it was also the recipient of substantial welfare entitlements. These entitlements enabled its members not only to escape from the status where "the only thing they could lose were their chains", but motivated them to enter various political alliances and further promote their interests in the existing framework.

Some Marxist theorists who attempted to describe these intermediary categories identified their special characteristics and recognized that they rightfully constitute a class of their own, the "middle class". Using the Marxist analytical framework, Antonio Gramsci (1891–1937), developed a theory of *organic intellectuals*. The growing numbers of professional executives is explained by the needs of production and their role is crucial for the credibility of the institutions and the function of the system. As Gramsci notes:

> The capitalist entrepreneur [creates alongside himself] the industrial technician, the specialist in political economy, the organizers of a new culture, of a new legal system, etc.. … The entrepreneur himself represents a higher level of social elaboration, already characterized by a certain directive and technical (i.e., intellectual) capacity: he must have a certain technical capacity, not only in the limited sphere of his activity and initiative but in other spheres as well, at least in those which are closest to economic production. He must be an organizer of masses of men; he must be an organizer of the 'confidence' of investors in his business, of the customers for his product, etc.[17]

This concept opened a debate on the social role of executives and professionals that continued throughout the twentieth century. The big question was whether, in the eventuality of a revolution, those experts would take the side of the working class or join the owners and strive for corporate expansion.

9.5.2 Evolutionary Theories

In the United States, it was Veblen, who first attempted to develop a theory in account for the new managerial class. Veblen saw the economy as a mechanical process where capital and labor are not the only important factors: entrepreneurship and technical knowledge were also crucial. Unskilled laborers were not the only ones being exploited, but business executives, the only ones capable of running the factories of capitalists, are also economically victimized.

The French philosopher August Comte took this analysis one step further, arguing that there are more internal divisions in the managerial class:

> Between the abstract theoretician and the floor managers of the manufacturing companies an intermediate class is being formed, the class of engineers. Its special purpose is to organize the relationship between theory and practice.[18]

[17] Gramsci, *The Intellectuals* (from Selections from the Prison Notebooks) (1971).

[18] Comte, *op.cit.*

Joseph Schumpeter (1883–1950) connected the managerial class with the new types of employment and promotion that accompanied the changes in the scale of production. In earlier societies, the pattern of heroic individuals held great social appeal to both the ruler and to the lower classes who were in desperate need of idealizing someone standing up to their master. In the first period of wild capitalism, the ideal of heroic opposition to those exploiting the workers retained its social appeal. But these ideals lost ground when more and more members of society were expecting to take part in the newly created wealth:

> Capitalist civilization is rationalistic and anti-heroic... Success in industry and commerce requires a lot of stamina, yet industrial and commercial activity is essentially unheroic in the knight's sense [...] No flourishing of swords about it, not much physical prowess, no chance to gallop the armored horse into the enemy, preferably a heretic or heathen—and the ideology that glorifies the idea of fighting for fighting's shake and of victory for victory's shake understandably withers in the office among all the columns of figures.[19]

9.5.3 Pareto's Social Hierarchy

Pareto made a thought-provoking contribution to the analysis of social power and the role of the middle class. As he did in his welfare analysis, Pareto again focused on the broader issue of status quo and whether it can be changed. He argued that amid all progress and economic development, there are certain power relations immune to change as they are entrenched by a phenomenon called '*the persistence of aggregates*'. To interpret low-scale shifts—what he called *residues*—he formulated a theory to explain how domination is anchored in human psychology and fostered by social institutions such as family, group settings and traditions. These institutions have a cumulative effect, entrenching the sentiments of deference, respect and fear, which are then used to cement social stratification and hierarchy.

Pareto asserts a social taxonomy with two major classes: one is the elite consisting of two sub-groups, the governing and non-governing elite; the other is the subjugated class.[20] Governing elites are constantly challenged by new ones, and can be replaced if the latter proved to be superior. Pareto thus replaced the Marxist taxonomy which defined social classes based on their role in production, with an "open system" of merit-based elite circulation. Hegemony cannot succeed even if based exclusively on the rational exercise of power but needs to be accompanied by actions of a psychological and symbolic character. Though constantly pursuing its own interests, the ruling class may improve the situation of the subject class.

According to Pareto, revolution is a violent change of the governing class, not of the conditions in production. By establishing his theory on psychological grounds, putting economic structures aside, Pareto provided legitimacy to social power

[19] Schumpeter, *Capitalism, Socialism and Democracy* (1942).

[20] Pareto, *The circulation of elites* (1961).

relations, under his own optimality principle: as long as all members of society accept their position, any change in social structures can be legitimate only if no one becomes worse-off. The subject class cannot improve its position in Pareto's system, unless the ruling class benefits too.

Pareto's theories became popular among those looking at the power dimensions of stratifications and professional status. Marcuse credits Pareto with being the first to analyze the psychological underpinnings of hegemonic relations in capitalism.[21] As Pareto's approach on social power relations is not moralistic or value-specific, Marcuse notes that it can be used both by the ruling and by the subject class alike.

The emergence of intermediate classes in the social hierarchy is attributed by Pareto not merely to material changes of the production process, but also to the opportunities of social advance that present themselves. Individuals belonging to the middle classes are not only increasing in numbers, but acquire important positions in the economic hierarchy and serve as models of social success throughout the Western World and especially so in the United States. According to this analysis, the middle class neither relates to the aspirations of Veblen to become the new governing class, nor to those of Gramsci who hoped to see it becoming an ally of the working class. They well stay neutral and seek self-improvent.

9.5.4 Neo-Marxist Revisionism

Following the Paretian framework, American sociologist C. Wright Mills brought the lofty expectations Marxists had harbored for the middle class to the realities of industrial society. Their intermediate position between capital and labor is the major reason behind their unwillingness to serve as a social vanguard. Using Marxist terminology, Wright Mills asserts that white-collars are:

> Estranged from community and society in a context of distrust and manipulation; alienated from work and, on the personality market, from self; expropriated of individual rationality and politically apathetic—these are the new little people, the unwilling vanguard of modern society.[22]

He then explains why nothing revolutionary should be expected from the middle class:

> ... But while practical politicians, still living in the ideological air of the nineteenth century, had paid little attention to the new middle class, theoreticians of the left have vigorously claimed the salaried employee as a potential proletarian, and theoreticians of the right and center have hailed him as a sign of the continuing bulk and vigor of the middle class. Stray heretics from both camps have even thought, from time to time, that the higher-ups of the white collar world might form a center of initiative for new political beginnings.

[21] Marcuse, *From Luther to Popper* (1983, pp. 150–155).

[22] See his book, *White Collar: The American Middle Classes* (2002, p. xviii).

Some Neo-Marxist economists moved beyond the concept of alienation and attributed the lack of class activism of white and blue collar workers in advanced countries to purely economic reasons. Arghiri Emmanuel (1920–2004), for example, developed the theory of "unequal exchange", according to which it is not only the capitalists who benefit from the exploitative wages and working conditions in underdeveloped countries. It is also the blue-collar and white-collar workers and executives in the developed world, who partially expropriate the surplus value through the mechanism of international trade and multinational production. Emmanuel attributed the absence of any significant solidarity shown by workers in the developed countries to their counterparts in the Third World, to the fact that the former get part of the surplus value produced in poor countries and thus they ascend to the middle class. As he explains,

> a de facto united front, or even a sense of solidarity, of the workers and capitalists of the well-to-do countries is directed against the poor nations. This front does not refute the traditional Marxist thesis concerning the internal antagonism between capital and labor; it only means that the struggle of the unions is now also about dividing the spoils.[23]

According to Emmanuel, workers in poor countries need to realize they if they wish to fight for independence, the struggle should be on their own; it is futile to expect solidarity—let alone salvation—from the working and middle classes of rich countries.

Two significant developments after World War II further enhanced the role and differentiation of the middle class. Firstly, strides in technology dramatically increased demand for middle executives, while new production methods curtailed demand for unskilled labor. The need of expansion in overseas markets led to the emergence of powerful multinational companies, which employed entire armies of skilled executives with more to say on the organization of production than merely reflecting class stereotypes. Secondly, crucial economic activities in many western economies came under the control of state-owned companies, with executives acquiring influence and power for their own sake thanks to lose oversight from the Government. Even in centrally planned economies, industries were managed by specialists who quickly replaced workers' committees and gradually evolved into a type of middle class.

9.6 Classical Economics: Heyday and Crisis, 1890–1930

Explaining the deficiencies occurring between supply and demand was one of the most challenging questions in economic thinking during the last part of nineteenth century. French economist Jean Batiste Say (1767–1832) had already tried to simplify the task by asserting that demand is solely determined by supply, thus

[23] Emmanuel, *Unequal Exchange* (1972).

demand-side shortages are ruled out and attention should focus on how to ensure the appropriate interplay of production factors.

Say's thesis (at that time it was called a 'law') was fiercely attacked by Marx who saw it as an excuse to further suppress wages in order to invigorate the motivation for profits. Other economists were equally uneasy and in search of a more convincing description of market dynamics. The major contribution to the understanding of the balancing mechanism between supply and demand was formulated by Léon Walras and later extended by Alfred Marshall.

9.6.1 The Perfect Equilibrium

Leon Walras (1834–1910), a French mathematician and economist, duly investigated how the market ensures the same price across consumers and whether this is unique and practically reachable.[24] By assuming N product markets, each one embedded with a function of consumers' demand and a function of producers' supply, he obtained a system of 2N equations with 2N unknowns in quantities and prices. Under specific conditions, he found that there can be only one solution balancing demand and supply in all markets.[25]

Walras was unable, however, to prove the existence and uniqueness of such a solution with mathematical vigor. So he conceived a theoretical model of economic behavior, which *could* converge to the solution with prices recursively adjusted until supply and demand are equalized. The process involves a hypothetical "auctioneer" who announces an initial market price for a specific product. If the price is lower than equilibrium price, consumers will increase their demand; since producers cannot keep up with new demand, prices increase, and so on and so forth. This mechanism of groping, or *tâtonnement* as it became known, is based on a continual process of trial and error that ultimately brings prices to equilibrium.

The self-equilibrating paradigm became a landmark in the history of economic thought. Economics could, at last, boast to be an exact science, as there was a strict foundation of individual choice to explain demand and supply, and also a unique solution could be reached even if it took some trial and error to get there. The intellectual triumph, however, was premised on some strong assumptions, including that:

(a) Perfect competition and perfect information reign in each market.
(b) Demand, supply and price adjustments happen immediately.
(c) No additional cost incurs in the transition from one state to another.

[24] Walras also focused on applied economic policy. In contrast to the physiocrats, he was in favor of taxing land rent, since he thought that the land's value comes from society and taxing it is a way to return part of this value to its source. Supporting industrial production, he was against taxing workers.

[25] The system is described in the book by Walras, *Elements of Pure Economics* (1954).

These preconditions are, of course, far from self-evident and may be violated, whenever respectively:

(a′) There are oligopolies distorting supply; consumers and/or producers are not perfectly informed or are endowed with asymmetric information.
(b′) Demand, supply and price adjustments are not immediate.
(c′) Transaction costs impede the transition to the point of perfect equilibrium.

Entire branches of modern economics deal with these problems, investigating oligopolistic markets, dynamic price adjustment and transaction costs. Despite its apparent lack of realism, the equilibrium model is still useful as a benchmark even if merely to evaluate the extent to which actual market operation is imperfect. Deviations from the model are thus seen as a measure of market failures and economists can analyze the benefits that could accrue by reducing or eliminating them.

9.6.2 Gradual Equilibrium

Building on Walras' achievement, new theories on market equilibrium emerged, this time based on more realistic hypotheses. Alfred Marshall (1842–1924) examined the gradual transition to equilibrium by adding the concept of time to the Walrasian model. Marshall explained why this price adjustment does not happen immediately due to problems of communication between producers and consumers or, in today's parlance, of partial information. Since prices do not adjust immediately, this may lead to excessive demand or supply, provoking temporary crises.

Convergence to equilibrium is only gradual. During transition, there is a state of partial equilibrium even if producers adjust their prices but wages remain rigid. This leaves certain factors (for example employment and investment levels) to remain in a state of imbalance for even longer periods, but ultimately fluctuations are eliminated. In the end, wages are also adjusted and investment keeps up with the new composition of production factors as determined by their marginal returns.

Marshall's *Principles of Economics* was very influential and remained the standard theory taught in economics departments for decades to come. It attributed the crises to inherent market factors and could explain why an economy may remain in a state of imbalance for some time. Still, Marshall ignored another eventuality: sometimes an economy needs a very long time to return to a state of equilibrium by itself and, by the time it does so, society has paid a heavy price in the form of high unemployment and general dissolution. Though Marshall believed that a certain price–wage adjustment may always prevent long term imbalance, it was one of his students, J. M. Keynes, who understood that such imbalances may turn out to be permanent unless decisive and timely intervention takes place.

9.6.3 The Role of Money

A major tenet of the new economic theories was the quantity theory of money. Irving Fischer developed the theory in its present form, and suggested that price stability rests exclusively on the amount of money in circulation without affecting real economic activity. This was the core belief of the school of economic thought that later came to be known as "monetarism".

Fischer argued that any price increase can be countered by limiting monetary circulation. This policy exerts upward pressure on nominal interest rates, so that real interest rates (which are calculated as the nominal rate minus inflation) remain unchanged. Since both economic activity and real interest rates do not change, the accumulation of capital is not influenced by fluctuations in nominal demand. According to Fischer, monetary policy can, even in a crisis, blissfully pursue its inflation-fighting policy without worrying about possible side-effects in investment or consumption, while any potential imbalance will be restored by a Marshallian adjustment. This approach failed miserably to account for and deal with the economic crisis that hit the American economy in 1893. Alas, this was not its last failure; the Great Depression and many more crises—and disappointments—were to follow.

With Marshallian adjustment and Fischer's monetary formula, economists believed that the whole process was figured out. Their premise, however, was a fragile one as it presumed that interest rates are of the same nature regardless of whether they apply to the banking system or the accumulation of capital. This confusion was solved by the Stockholm School which left its own mark on the evolution of economic thought. An impact well worth the effort, as it turned out to be.

9.6.4 The Wicksellian Wedge

Swedish economist Knut Wicksell (1851–1926) developed a theory, radically different from the Fischerian approach, in his quest to identify the factors influencing investors' behavior and the role of interest rates in the business cycle. According to Wicksell, an investment is an act which aims at some change in production that will bring returns in the future, and needs equally long-term savings to be funded. Though balancing eventually, the two processes are more likely to evolve differently in the meanwhile.

To explain the divergence, Wicksell asserted that there are two types of interest rates respectively: one, called *natural rate of interest*, refers to the return a consumer expects to get from his savings; the other, called *money rate of interest*, is employed by the banks to charge businesses for credit provision.[26] Most of the time

[26] Wicksell wrote about this at (1898) "The Influence of the Rate of Interest on Commodity Prices"; see Lindahl, *Selected Papers on Economic Theory by Knut Wicksell* (1958, pp. 67–92).

the two interest rates are not identical, and the reason is that savings and business credit are determined by different mechanisms. Two cases are identified:

(a) If the money rate of interest is lower than the natural rate, banks issue more loans, thus boosting investment levels and real economic activity. If consumers fail to respond to this boom, the result is a crisis of insufficient demand.
(b) If the money rate of interest is higher than the natural rate, investments plunge, businesses shut down and production enters a recession. If demand does not adjust quickly to the new conditions, a supply-side crisis ensues.

In an attempt to understand the mechanism setting a business cycle off, the Stockholm School focused on the role of expectations held by market participants on the natural rate of interest, and by doing so another issue was introduced in the agenda of open questions, foreshadowing the debates in late twentieth century.

The findings and conclusions of the Swedish economists were contested by other schools of economic thought. The Austrian school, for example, dismissed the idea that monetary expansion can be used to cut the money rate of interest, invoking the danger of inflation and the uncertainty it would cause concerning the real return on investments. Both sides were stuck in their positions for a long debate that would ensue in the Great Crisis in the 1930s. At that point, Keynes would echo much of the Wicksellian insights by proposing a policy of adjusting the money rate of interest as the means to encourage investment and help the economy escape recession.

References

Barro RJ (1974) Are government bonds net wealth? J Polit Econ 82(6):1095–1117
Comte A (1971) A general view of positivism (Discours sur l'ensemble du positivism). Translated from the French by J.H. Bridges. Dubuque (Iowa): Brown
Emmanuel A (1972) Unequal exchange. A study of the imperialism of trade. Monthly Review, London
Gramsci A (1971) The intellectuals (from Selections from the Prison Notebooks) of Antonio Gramsci (edited and trans: Quinton Hoare and Geoffrey Nowell Smith). International Publishers
Hume D (1748) Of the original Contract. Available at http://www.constitution.org/dh/origcont.htm
Johansson P (1991) An introduction to modern welfare economics. Cambridge University Press, Cambridge, UK
Katz M (2013) The undeserving poor: America's enduring confrontation with poverty. Oxford University Press, Oxford
Kitromilides P (1998) Political thinkers of the modern era: biographical and interpretative essays. Poreia Publications, Athens (in Greek)
Marcuse H (1983) From Luther to Popper. Verso, London
Mill JS (1981) On liberty. Penguin, London
Nicholas D (1992) The evolution of the medieval world: society, government and thought in Europe 312–1500. Longman, London
Olson M (1965) The logic of collective action: public goods and the theory of groups. Harvard University Press, Mass

Pareto V (1961) The circulation of elites. In: Parson T, Shils E, Naegel K, Pitts J (eds) Theories of society: foundations of modern sociological theory. The Free Press of Glencoe, New York, pp 551–557
Popper K (1999) All life is problem solving. Routledge, New York
Rawls J (1999) A theory of justice. Harvard University Press, Cambridge, MA
Sandel M (2009) Justice: what's the right thing to do? Farrar, Strauss & Giroux, New York
Schumpeter J (1942) Capitalism, socialism and democracy. Harper & Row, New York
Walras L (1954) Elements of pure economics (trans: Jaffes W). Routledge, London
Wicksell K (1898) The influence of the rate of interest on commodity prices. In: Lindahl E (ed) Selected papers on economic theory by Knut Wicksell (1969). Augustus M Kelley, New York
Wright-Mills C (2002) White collar: the American middle classes. Oxford University Press, Oxford

The Great Crisis and the Theory of Keynes 10

Abstract

No other single event has so profoundly transformed the fundamentals of economic thinking as the Great Crash in 1929. In the aftermath of the crisis, academic orthodoxy and policy makers were passively waiting for the economy to automatic return to normality, while Marxist orthodoxy was contemplating the doomsday of capitalism. It took a radical new approach to understand the intricacies of the crisis and mobilize a new type of policies that ultimately succeeded in restoring employment and confidence in the system. The rise of Keynesianism inspired a host of related disciplines on measuring and modeling economic phenomena and ushered in a new period of understanding economics and conducting economic policy. Studying fluctuations became a central theme in economic theory, but views were still in conflict as to whether they can be dampened by policy intervention or should be left to attenuate themselves.

10.1 The Great Crisis in 1929

In early twentieth century, the hitherto deadliest war in history took place. Belligerent nations abandoned the Gold Standard, so that they could run the printing presses to fund war spending without fear of massive demand to convert currencies into Gold. Inevitably, most economies plunged into recession and suffered from high inflation after World War I; facing a loss of competitiveness, governments devalued their currencies, but this policy further fueled inflation pressures.

To avoid a currency war that would lead to mutual economic annihilation and hyperinflation, the United States decided to adopt the Gold Standard in 1919. Three years later, the European powers organized the Geneva Convention to address how to ensure stability in the foreign exchange markets. Participating nations hoped that this would bring domestic inflation down and lower interest rates in international bond markets, stimulating investment and growth in their war-torn economies. Around 20 countries immediately adopted the standard, and Britain followed suit

N. Christodoulakis, *How Crises Shaped Economic Ideas and Policies*,
DOI 10.1007/978-3-319-16871-5_10

in 1925. To reaffirm continuity, Britain even kept the pre-war exchange rate of pound sterling to gold, a decision soon to be followed by mounting pressure on its external balance. The British pound was also dethroned as a global currency after the War, and the United States was now the state with the amplest gold reserves. Therefore, it was the American dollar that emerged as the new "anchor" of the international system.

This meant that all other countries had to maintain a stable exchange rate to the US dollar; in order to achieve that, they had to mirror the monetary policy followed by the United States. Pegging their currencies to the dollar was very efficient and inflation fell rapidly. Ten years later, however, this system would be transformed from an "anchor of stability" into a crisis-transmission mechanism. The Gold Standard collapsed when policy makers were no longer willing to play along.

10.1.1 The Crash of 1929 in the US

Since 1921 the New York Stock Exchange experienced a period of soaring prices, with the index rising ninefold, from 50 to 450 points. Interest rates went too high and eventually a crisis erupted in October 1929, when the NYSE crashed. It would soon spread to other sectors of the US and the international economy; production took a dive, while unemployment skyrocketed, severely hurting the disposable income of households. As demand fell, prices followed suit, with the economy entering a period of severe deflation. The crisis had gone on for 4 years (1929–1933) before the economy entered a path of gradual recovery. Three phases of the crisis are distinguished[1]:

Phase I: In 1930, bank runs occurred in several American states and soon an epidemic of fear led to the collapse of hundreds of banks. A second bank crisis occurred in 1931, but it had been the collapse of banking giants in Austria and Germany that impacted two different effects on the United States: on the positive side, European capital was in need of a safe haven; on the other hand, many US banks had to write down investments in European banks causing American depositors to panic once again, massively withdrawing savings and depleting the liquidity of the system.

Phase II: Depletion of gold reserves due to the convertibility rule. In 1931, Britain abandoned the Gold Standard, sending shock waves through the markets with investors fearing that the US would soon follow suit. Both foreign and domestic holders of dollars rushed to convert them into gold. The double drainage on gold reserves restricted money supply, liquidity plunged and the real economy was facing an even worse recession.

Phase III: The third panic occurred in 1933 in the financial sector as yet another bank run was caused by the persistent fear of imminent bank collapses. Bank holidays and the imposition of limits on payments were two of the measures

[1] More details can be found in Friedman and Schwartz, *The Great Contraction 1929–1933* (2008).

authorities took to prevent a liquidity meltdown, but this led depositors to hoard more cash. In the meantime, demand for gold and foreign currencies skyrocketed during the election campaign of 1932, as there was widespread fear that the new administration would devalue the dollar. This fear further squeezed money circulation and economic activity, aggravating the recession to the point of social disintegration.

These consecutive panic-stricken and recession-fuelling developments were not yet seen at the time as a general meltdown process, thus policy makers did not consider it necessary to respond with an equally decisive counter-attack. Responses were ad hoc and often contradictory in nature. To wit, when the crisis of 1929 erupted, the central bank stubbornly kept interest rates high in an attempt to prevent bank runs, but this only aggravated the contraction in the real economy. With prices dropping due to anemic demand, the real debt of businesses and households increased since they were compelled to save a larger part of disposable income to service the burden. This naturally depressed demand even more, prices fell further, and the economy entered a vicious circle of recession and growing indebtedness. Many businesses ultimately shut down, and unemployment peaked.

10.1.2 Missing Theories

During the crisis, which was marked by periods of deterioration and brief respite, there was a glaring inadequacy of theories explaining its causes and characteristics. No comprehensive policy response, that would allow the economy to escape the crisis, was forthcoming. The US government had initially decided to bolster public spending, in order to stimulate demand, but this was ditched in 1931 as the view that high budget deficits would be even more catastrophic prevailed. In fact, the US government imposed tax increases to compensate for the increased spending on unemployment benefits and to balance the budget of 1932. Rising taxes naturally depressed demand even further, pushing the economy to new all-time troughs.

As difficult as it may be to understand this self-contradictory policy today, back then it had many supporters; in fact, approval was almost unanimous as most economists of the time were under the illusion that fiscal tightening would automatically restore confidence in the economy, diffusing the panic.[2] Even Franklin Roosevelt, who would pursue a radically different policy after 1932, did not dare to stand up against the thus far dominant orthodoxy of fiscal tightening.[3] Only when recession became abysmal, did policy makers start looking for a new prescription.

The collapse of the American economy quickly evolved into an international crisis, spreading through the rigid exchange rates mechanism. When a European

[2] The illusion was fueled by the groupthink effect: the pressure to conform with the prevailing view discourages and punishes individual dissent.

[3] As Friedman and Schwartz observe, during the 1932 campaign, both presidential candidates had promised to balance the budget if elected (2008, *op. cit.*, p. 48).

country abandoned the Gold Standard, its currency devalued dramatically and its economy was becoming more competitive. This meant that it was only a matter of time before other nations would join the chorus and this led to serial devaluations and inflation. The Gold Standard met an inglorious and premature end just before nations began preparing for the next bloodshed of World War II.

10.2 Conflicting Interpretations of the Great Crisis

10.2.1 Marxist Contemplation

All major theoretical approaches tried to explain the crisis from their own point of view. Marxist economists saw the crisis as the obvious confirmation of their theories and the prelude to the mush awaited socialist transformation. According to Eugene Varga, chief economist of the Communist International, the crisis:

> caused a profound disturbance of the entire capitalist system, initiated a new and higher stage of the general crisis of capitalism and resulted in the maturing of the objective pre-requisites for the revolutionary crisis.[4]

Comparing the crumbling economy in the US with the fast industrialization in the Soviet Union, produced a wave of arguments in favor of central planning and seemed to confirm the belief that crises would be automatically eradicated under socialism.

10.2.2 Liberal Counselling

According to liberal economists, led by Milton Friedman (NLE 1976), the economic collapse was the result of a mistaken monetary policy and the unwillingness of Central Banks to act as "lenders of last resort". Had they met the capital needs of the banking system and regulated the circulation of money in a way that would calm panicked depositors, the crisis would have been avoided. In his remarkable book, *A Monetary History of the United States*, Friedman analyzes many banking crises that had occurred during the nineteenth and early twentieth century and makes clear how crucial the role of monetary policy is, paving the way for the theory of monetarism.

The Austrian School, with von Misses and Hayek in the frontline, put forward an even more radical policy prescription. They argued that monetary intervention should have been avoided at all costs and the economy should be left alone, for the cure to come through the unhindered operation of markets. The Austrian School dismissed monetary easing, lest lower interest rates wreak havoc in the process of

[4] From Varga, *The Great Crisis and Its Political Consequences: Economics and Politics 1928–1934* (1935). Quoted from Howard and King, *A History of Marxian Economics: 1929–1990 vol. II* (1992, p. 5).

capital accumulation. Friedman's theory is not as radical as the Austrian school's, but their views are identical on two crucial points: they both defend the free markets and ignore the role of demand management in restoring trust and confidence in the economy.

Similarly, Fisher believed that the business cycle is short-lived and any intervention in the fundamentals of the economy was therefore unnecessary, if not outright harmful. But at least, he acknowledged the problem of high debt in a deflationary environment arguing in favor of policies aiming at stimulating economic activity and preventing price falls. His famous phrase '*The more the debtors pay, the more they owe*' was a warning on why contractionary behavior and policy during a crisis may lead to exactly opposite effects than those pursued; Fisher (1933). In this way, Fisher influenced economic thinking to move toward the new paradigm of Keynesianism.

The most important English economists (headed by Arthur Pigou at Cambridge University) never lost confidence in the Marshallian adjustment model. According to them, balance would have been restored and unemployment decrease, had wage levels dropped more rapidly.[5] Pigou attributed the crisis to the fact that nominal wages did not fall proportionally to price levels, thus implying that recession and unemployment were caused by the increase in real wages.

10.2.3 The Emergence of the Keynesian Theory

A radically different account of the crisis was provided by a new school of economic thought that soon would become known as Keynesianism, after its most prominent proponent, John M. Keynes (1883–1946). In this account, the single most important cause of the Great Depression was the collapse of aggregate demand, as households lost confidence in the economy. Fearing a prolonged economic downturn, they hoarded cash instead of consuming or even saving their income in banks. Bank liquidity was depleted and the economy asphyxiated, while recession and unemployment went from bad to worse, intensifying hoarding behavior. This led to the paradox of thrift: savings may have been a prudent choice at the level of individual households, but if everyone tries to save more money, aggregate demand will fall and recession will worsen.[6]

According to the new theory, developed in Keynes's (1936) seminal book *The General Theory of Employment, Interest and Money*, the proper response is a decisive bolstering of demand, even on a massive scale if necessary. At such a magnitude, the demand-side stimulus can be delivered only by the government, which should increase public spending to fight unemployment and support low-income families. Households will then spend the extra money, and businesses will be convinced that

[5] In his book Pigou, *The Theory of Unemployment* (1933).

[6] The analogy looks impressively similar to the distinction initially made by Machiavelli between individual preferences and collective outcomes, as discussed in Chap. 4.

investing is again a safe choice so that gradually the economy returns to full activity. Other economists, working on the same issues, also contributed to this theory. A landmark contribution was the concept of the "multiplier", introduced by Robert Kahn, describing the mechanism by which a demand stimulus leads to an even larger increase of economic activity.

10.2.4 The Cambridge Precursors

Another Cambridge economist dealing with these issues was Michal Kalecki (1899–1970); a Polish Marxist and among the first who noticed the importance of the economic multiplier. In Kalecki's model, workers spend their entire wages on consumption, so the multiplier is significant. Any increase of their disposable income will be channeled almost exclusively to consumption, bolstering production and employment provided that there are idle capital resources. Thus, one way to avoid economic turbulence is to use the multiplier effect either by increasing workers' wages, since they will immediately consume the extra income, or by government spending, which can make up for the lack of private investment. Higher employment then further increases demand, allowing the economy to stabilize at higher levels of activity.

On the other hand, any increase in the demand by capitalists will usually end up in purchases from other capitalists. There is therefore no positive outcome and aggregate capitalist savings do not change. Kalecki summarized this phenomenon in his famous phrase "workers spend what they gain, while capitalists gain what they spend". He was the first to criticize saving behavior during a crisis. According to Kalecki, deposits should decrease during a crisis, so that the economy can be kicked back into motion.

Building a multiplier-accelerator model, Kalecki developed a theory of business cycles with fluctuations occurring when the propensities to consume and/or invest are too low.[7] Since profits are proportionate to the capital employed, profit expectations further encouraged fixed investment by businesses. If fixed investment and production increase at the same rate, any new investment will increase production even further, accelerating activity, thus a virtuous cycle of increased production and consumption follows. If fixed investment expenditure is insufficient, however, production will stagnate, demand will drop and investment will plunge, unless the government intervenes and takes up the initiative.[8]

Another heated debate concerned the quantity theory of money, which was no more considered sufficient to interpret the behavior of households during a crisis.

[7] Kalecki, *Theory of Economic Dynamics* (1965), and Kalecki, *The Dynamics of the Capitalist Economy* (1977).

[8] Ricardo had already noted that "there cannot... be accumulated in a country any amount of capital which cannot be employed productively"; in his work *On the Principles of Political Economy and Taxation* (1817, Chap. XXI).

The "Cambridge equation", as it became known, drastically changed the focus of the quantity theory by examining the factors determining money demand during turbulent times. Economists associated with the University of Cambridge developed the concept of 'liquidity preferences', which encapsulates the desire of holding cash. Real balances correlate positively with the value of transactions and negatively with interest rates which represent the opportunity cost of holding cash. Their original contribution was the idea that holding cash becomes preferable in times of uncertainty, for example when individuals fear a collapse of the banking system. The greater the fear, the more powerful the liquidity preference becomes. If real money supply remains stable under such conditions, the economy inevitably contracts. To avoid this outcome, the central bank must restore the credibility of the financial system and intervene, either by cutting interest rates and/or by increasing money supply.

10.2.5 The Illusion of Automatic Rebalancing

During the post-war period, Keynesianism became so extensively dominant, that it was often taught as a self-evident dogma. It was *the theory*, the only game in town. Gradually, however, it lost sight of the original conditions that had led to its culmination and ended up being taught as a simplistic method of policy fine-tuning during a recession. Most economic literature of the time was based on crude simplifications that obscured even the causes that had led to the emergence of Keynesianism in the first place. The central idea of uncertainty was mostly forgotten, as the world was experiencing a prolonged period of postwar economic stability. Comparing the prevailing paradigm to the original insights, Cambridge economist Joan Robinson (1903–1983) branded it as "bastard Keynesianism". Some of the dangers lurking in this self-assured version of Keynesianism are the following:

To start with, Keynes had developed his theory starting from the fundamental observation that an economy spends longer periods in a state of imbalance rather than balance. Mechanistic "Keynesian" teaching, however, guided many generations of economists, into blindly believing that all an economy needs to return to equilibrium is the coordination of fiscal and monetary policies. They never examined whether this new equilibrium is viable or how this could be achieved in case it isn't.

A second misunderstanding of Keynesian theory resulted from confusion as to when two phenomena are in a state of balance and equilibrium. Some phenomena are not always equal or of the same nature, but eventually they converge to the same level. Such confusion frequently occurs when equating investments with savings by the famous textbook expression $I = S$. Savings and investment are typically analyzed as increasing and decreasing functions of interest rates respectively. Far from being an identity, this relation describes the *final* outcome of a gradual process of savings and investment decisions, as Wicksell's intuitive theory had demonstrated long before Keynes; see Sect. 9.6.4.

Actually, not even the interest rate which affects both sides of the equation is the same. In the right-hand side of the equation it represents the return (i) of money deposits, which implies a rising function $S = S(i)$. The interest rate in the left-hand side represents the cost of capital investment (r), implying a downward function $I = I(r)$. Essentially, (r) is a monotonic function of bank interest rates (i), but the two are equalized only in conditions of perfect competition and substitutability of financial and physical assets. The chance that all these conditions are met or can be kept applied for a prolonged period of time is rather slim.

Acknowledging this difference, allows us to gain a deeper understanding of crisis dynamics. During crisis, households suffer or feel threatened by unemployment, so they preemptively increase their savings to deal with future needs. As consumption drops, businesses decrease investment and focus on maintaining high liquidity levels so as to deal with an eventual credit crunch. The two levels of savings and investment are now in a state of imbalance. Post crisis, savings drop as households, no longer fearful of unemployment, spend more. Investment also increases, to satisfy rising demand, so that balance gradually returns.

Analyzing this phenomenon—and having benefited from the insights of the Swedish School—Gunnar Myrdal (NLE, 1974) identified the distinction between *ex ante* (or planned) and *ex post* (or realized) savings and investment, describing their behavior during the business cycle: when investment increases, production, incomes and savings follow suit and this eventually leads to $I(r) = S(i)$. During a crisis, savings initially increase and consumption drops. As investment, production and incomes fall as well, the downward spiral leads to lower savings, which are again equalized with lower investment.

10.3 Measuring the Economy and the Policy Effects

After WW II, all streams of economic theories vouched to get quantified. The prevalence of the new Keynesian paradigm meant that fresh tools of economic representation and analysis were needed to quantify policy interventions prescribed by the theory. This marks a new era of economic modeling to an extent of elaboration and application hitherto unknown. Previously, the only available framework was the Walrasian system extended by Marshall to take price rigidity into account as examined in Sect. 9.6.2.

10.3.1 National Accounts

The use of government budgeting had already become widespread in the 1930s, providing a coherent framework of national accounts and allowing a better understanding of policy effects. As has happened with many other scientific discoveries in history, the need for systematic economic measurement and record-keeping became more pressing in the course of the two World Wars. After the end of the Great War, the US army had suffered significant losses partly due to insufficient logistics, as war supplies and provisions were not optimally allocated. The US government decided to improve military logistics, assigning the task to such economists as Simon Kuznets and Theodore Schulz (NLE 1971 and 1979 respectively). In his seminal work *National Income and Its Composition 1919–1938*, Kuznets laid the foundation for the modern methodology of calculating GDP.

Britain took similar steps, appointing economists such as James Meade and Richard Stone (NLE 1977 and 1984 respectively) to take charge of competent departments of the War Office. Their task was to organize government budgeting in such a way as to ensure the efficiency of logistics and war spending. The two economists used the Keynesian theory of aggregate demand to build a National Accounts model, thereby creating a framework capable of portraying and analyzing economic phenomena at a large scale.

Other quantitative frameworks also emerged during that time. General Equilibrium Models attempted to describe the whole specter of economic activity, including government policies. In 1946, Wesley Mitchell and Arthur Burns co-authored a report on *Measuring Business Cycles* where major statistical series of the American economy are classified and analyzed without adopting any specific economic theory.

Complicated models also emerged in the centrally planned economies, mainly in the Soviet Union. A different approach was developed by Wassily Leontief (NLE 1973) who introduced the input–output table, based on the idea of Quesnay's *Tableau*. The input–output model was adopted by many diverse economies—from the Soviet Union to the United States, to Netherlands and France—as a method to analyze the interdependencies between different economic sectors and monitor the balance between supply and demand.[9] Even today, international organizations, such as the United Nations, use international trade flow tables to represent imports and exports globally.

10.3.2 Estimated Models

The twentieth century saw the development of econometric models, describing economic relations by estimating statistical data. Norwegian economist Ragnar Frish (1895–1973) was the first to anticipate the importance and application of such tools from early on. Frish described the economy as a dynamic system which he subsequently solved to study business cycles and the existence of steady states. Working independently on the same problem, Dutch economist Jan Tinbergen (1903–1994) solved the system of estimated equations and suggested that this model can be used in forecasting and policy evaluations. Jointly Frish and Tinbergen were the first to win the Nobel Prize in Economics established in 1969.

[9] For a wide range of applications, see Leontief, *Essays in Economics: Theories, Facts and Policies* (1977).

Economists saw the merit on following the path opened by Frisch and sought for models to account for the fluctuations of the business cycle. The analysis of dynamic economic models with multiple equations demanded, however, increased computing power. Progress was slow until the 1970s when computers spread widely, opening up new horizons for economic researchers. Lawrence Klein (NLE 1980) is the father of economic model forecasting. He developed an econometric model for the American economy, featuring dozens of equations, and solved with the aid of high-powered computers. His methodology soon became popular worldwide, and almost all national economies built their own version of macroeconomic model based on Klein's archetypal. The globalization of the modeling tool soon transfigured the Keynesian premises on which the model was based to a sort of universal prescription on how an economy should be managed, adding a further layer of uniformity in the postwar economic landscape.[10] Uncertainty and shocks were benign and in any case they could be dealt with by using the elaborate techniques of the "optimal control theory" devised in the early 1950s by the great mathematician Richard Bellman.

Most of the models described above have by now fallen into disuse. The model of central economic planning, for example, has been abandoned by almost every nation, while national development plans are adopted on a case by case basis, as, for example, in the EU to assess the effects of cohesion policies. Large scale econometric models experienced their heyday 30 years ago, but since then their credibility has been challenged and are now used more sparingly.

The meteoric rise and fall of econometric models is not a chance event. When economic crises occur, they do not undermine just the dominant theoretical paradigms, but also their econometric and analytical frameworks. In his analysis of the history of science, Thomas S. Khun argues that the emergence of new theories demands "large-scale paradigm destruction and major shifts in the problems and techniques of normal science".[11] The emergence of a new economic theory led to an explosion of new techniques trying to validate and establish its credibility. In its heyday, this outburst often popularized and accelerated the dissemination and acceptance of the new theory—until it ultimately faced its own limitations and was superseded by new patterns.

[10] Time-series autoregressive models was another method of economic forecasting developed by Robert Engle and Clive Granger (NLE 1987). Instead of Keynesian principles and structural equations, it was based on the properties of time series without adhering on any specific theory.

[11] Khun, *The Structure of Scientific Revolutions* (1970, Chap. VII). He illustrates his point with the example of the series of mistaken forecasts made by the Ptolemaic model, which led to its ultimate replacement by the Copernican sun-centered model.

10.4 Business Cycle Theories

The study and analysis of business cycles has always posed huge challenges to economic theories. Early attempts were often fanciful. Jevons, for example, attempted to interpret fluctuations of agricultural productivity by invoking the cyclical behavior of the solar system.[12] He argued that sunspots cause climate alterations which cyclically affect rainfall and by extension agricultural production and economic activity. During the twentieth century, advances in econometrics allowed a more systematic research on business cycles by measuring their duration, timing, intensity and causes.

10.4.1 Keynesian Multipliers

Unsurprisingly, Keynesian theory prevailed for some time as the dominant explanation. Its proponents argued that economic crises are caused by insufficient demand, which, via a negative multiplier, leads the economy into recession. Economic uncertainty and the lack of confidence imply that investors cannot be counted upon to automatically lead the economy out of the crisis. Hence, it is the duty of the government to intervene and restore trust, bringing the economy back on a growth path and paving the way for the revival of investments.

In the post WWII era, Government intervention became the standard tool to deal with the consequences of crisis not only in the Anglo-Saxon world but almost everywhere, regardless of geography or the prevailing political system. Similar applications can be found in the inter-war period as well. For example, in 1936 France the Popular Front, an alliance of socialists and communists, increased wages to bolster demand and established many workers' rights. In Italy, Mussolini's fascist regime proceeded towards extensive nationalizations of banks and businesses. In Sweden, Social democratic governments financed public work projects to address the rising unemployment caused by the crisis.[13] In Hitler's Germany, Hjalmar Schacht, Minister of economics and central banker, gave a big push to industrial production, achieving spectacular growth rates and low unemployment.

This overarching acceptance of government's role in the economy cannot, however, conceal the great diversity of views adopted by various countries with regard to the operation of markets, the free allocation of production factors, as well as the exercise of freedom and political rights in general. Nations which violently curtailed such rights (Germany, Italy) collapsed after a deadly war; nations where the state assumed total control of the economy (Soviet Union and Eastern Europe) did last longer but ultimately collapsed, albeit peacefully. On the other hand, nations that followed Keynesian policies firmly ground on economic and political

[12] See Pressman's account in *Fifty Major Economists* (2006, p. 88).

[13] Gunnar Myrdal was one of the leading proponents of this policy.

freedom (such as those in Western Europe) witnessed a rapid growth of their economies and were better able to cope with disruptive economic crises without collapsing.

Even these economies, however, could not cope with a shock of a nature and magnitude previously unknown. The Keynesian interpretation of the crisis ignored the role of supply-side factors and failed to explain the effects oil price shocks had on western economies in the 1970s, as will be discussed in Chap. 12.

10.4.2 The Theory of Creative Destruction

Alternative theories attempted to better explain the consequences of economic crisis not only on demand but also with regard to the restructuring of the business world. Joseph Schumpeter, an admirer of Walrasian theories, saw economic crises as an inevitable process of adaptation to a changing environment. By combining the concepts of (a) rising returns as a result of the division of labor described by Adam Smith, and (b) diminishing returns of the factors of production described by Marx, Schumpeter identified the pressures which firms constantly face for innovation and new technologies.

Economic crises were viewed as a purging mechanism which destroys those who are unable to adjust, forcing many businesses to shut down. The lay-offs that follow drive unemployment up, and wages fall—as Marx had predicted. But instead of the ultimate collapse of capitalism, Schumpeter anticipated that the economy enters a period of creative restructuring. Taking advantage of lower wages, some firms recover and merge into bigger conglomerates. Thanks to their bigger size, they are better equipped to innovate and improve their competitiveness. During recovery, more jobs are created, unemployment falls, wages rise, and the economy enters the growth stage of the business cycle. In several other occasions, theories of creative destruction were no more than truisms, since always something emerges successful after a major catastrophe.[14]

10.4.3 The Long Waves

Soviet economist Nicolai Kondratiev (1892–1938) attempted to connect the occurrence of wars with the various phases of a long-term business cycle, averaging 50 years. According to his theory, the expansionary phase of the cycle presages war, as nations realize that economic activity will soon decelerate due to diminishing returns. Therefore, governments search for new spheres of influence and profits so as to escape the next two phases of the cycle: stagnation and recession. During the

[14] Schumpeter failed monumentally to understand the impact of the Great Crash and believed that the recovery will automatically come; see Galbraith, *History of Economics: The Past as the Present* (1998, p. 195).

declining phase of the cycle, social groups react to the forthcoming recession, trying to entrench their interests. This causes great social upheavals every 50 years, with external wars erupting approximately every 25 years.[15]

As a matter of fact, European history offers many episodes vindicating long waves. Take for example the Seven Years' War between Anglo-Prussian and Austrian-French camps ended in 1763. The French Revolution took place in 1789 (26 years later), while the Napoleonic Wars continued until 1815 (52 years after the end of the Seven Years' War). In 1848, social revolt erupts in France (33 years after 1815). In 1890, the Crimean War began, while 1917 is the year of the Russian Revolution (27 years later). The end of World War II comes in 1945, while 1968 is the year of the French May and similar uprisings in Germany and Czechoslovakia (23 years later).

At first sight, the cycles seem to confirm Kondratiev's theory. But as no inherent explanation is offered, it is mere guesswork if one attempts to forecast when the next social explosion or war will take place based on such findings. For these are historical coincidences rationalized *ex post*, and may or may not have anything to do with social behaviors: during the expansionary phase of the cycle, a nation may indeed think that a conquest overseas will secure vital space. It is entirely possible, however, that the same nation instead decides to invest in technology as a way to deal with diminishing returns.

During the decline of the cycle, a conflict between social classes might indeed become inevitable if the weaker social classes are subject to exhaustion in order that ruling elites remain unscathed. Again, it is entirely possible that, instead of a strife being escalated, a process of creative destruction takes place, as Schumpeter's theory predicts, or that the government intervenes to alleviate the consequences of the crisis, as Keynes' theory proposed.

References

Barnett V (1998) Kondratiev and the dynamics of economic development. Macmillan, London

Fisher I (1933) The debt-deflation theory of great depressions. Econometrica 1(3):337–357

Friedman M, Schwartz A (2008) The great contraction 1929–1933. Princeton University Press, Princeton

Galbraith JK (1998) History of economics: the past as the present. Penguin, London

Howard M, King J (1992) A history of Marxian economics: 1929–1990, vol II. Macmillan, London

Kalecki M (1965) Theory of economic dynamics: an essay on cyclical and long-run changes in capitalist economy. Kelley, New York

[15] An account of his life and theory can be found in Barnett, *Kondratiev and the Dynamics of Economic Development* (1998). Despite being a Marxist, Kondratiev did not hesitate to argue that diminishing capital returns would exist under socialism as well, thus cyclical crises would occur despite state ownership of all means of production. This view was an arrow in the heart of established certainty and he was exiled to Siberia. Kondratiev never denounced his views and ultimately was executed by Stalin in 1938. He was reinstated in 1987.

Kalecki M (1977) Selected essays on the dynamics of the capitalist economy 1933–1970. Cambridge University Press, Cambridge
Keynes JM (1936) The general theory of employment, interest and money. Macmillan, London
Kuhn T (1970) The structure of scientific revolutions. University of Chicago Press, Chicago
Leontief W (1977) Essays in economics: theories, facts and policies. Blackwell, Oxford
Pigou A (1933) The theory of unemployment. Macmillan, London
Pressman S (2006) Fifty major economists. Routledge, London
Ricardo D (1817) On the principles of political economy and taxation. John Murray, London
Varga E (1935) The great crisis and its political consequences: economics and politics 1928–1934. Modern Books, London

Theories of Central Planning and the Socialist Crises

11

Abstract

Marxist theory was certain about the inevitable demise of capitalism but less so on the system that would succeed it. Critical problems of production and distribution were only coarsely dealt with, thus the practical implementation of socialism was full of ideological experimentation, policy shortcomings, and food shortages. Early famines led to the formation of a highly bureaucratic central planning which enabled a speedy industrialization of Soviet Union but failed in the production of consumer goods and the efficient allocation of investments. The failures of central planning invited a variety of critiques, from the suggestion of an intermediate mix of markets with socialist production to the utter denouncement of price dirigisme in the economy. Several economic and political thinkers put in doubt the very character of the socialist system and tried to explain it as a modern replica of Asiatic despotism.

11.1 Marxism and Central Planning

Before examining the effectiveness and the legacy of central planning, it is fair to lay the immense challenges put forward in practice:

11.1.1 Unchartered Waters

The first question revolved around which social class should take the lead in overturning the capitalist system. In the era of classical economics, there were clear lines separating social classes and it was easy for theorists to identify with, and defend, a specific group (as Adam Smith had defended producers, Colbert the merchants, physiocrats the farmers and so on) or oppose it (as Ricardo castigated the landlords or the Church had denounced commerce in Middle Ages). A deep clear-cut class division is also present in the theory of Marx and Engels. Writing in

N. Christodoulakis, *How Crises Shaped Economic Ideas and Policies*,
DOI 10.1007/978-3-319-16871-5_11

an environment of intense social polarization brought about by nineteenth-century capitalism, they unequivocally defended the working class and condemned its exploitation by the private owners of production.

According to their analysis however, Russia was not yet ready to undergo a transition to communism, lacking a working class mature enough to take the lead.[1] The Russian economy had to complete its course along the path from primitive accumulation to the stage of advanced capitalism which would be pregnant with its unavoidable destruction. Lenin addressed this issue by arguing that the deficiency of history could be compensated by the action of a political party that would act instead, and on behalf, of the missing working class.[2]

It is true that at least in the original Marxist theories, the abolition of capitalism was not unavoidably connected with the establishment of powerful central authority. According to some approaches, this was far from being a necessity. For example, Georges Sorel (1847–1922), a theorist of revolutionary syndicalism, dismissed the emergence of some new ruling elite that would dominate the proletariat, and proposed instead the establishment of a loose federal system of worker communities, without an official hierarchy.[3] Marcuse heavily criticized this idea as an illusion and accused Sorrel of adopting a static view of power and failing to appreciate the dialectic of social dynamics.[4] He also argued that the need for a disciplined, highly productive labor force in socialism would inescapably lead to the emergence of "revolutionary elites", much like Lenin's party avant-garde. Marcuse from early on asserted that those aloof, self-righteous groups would become genuinely authoritarian elites, and actual history would just vindicate his prognosis.

11.1.2 Missing Markets

Concerning exchange and distribution, markets were already dismissed by the founders in the belief that all the crises and alienation caused by capitalist markets would disappear along with them. This would be automatically effected since

> [w]ith the abolition of the basis of private property, with the communistic regulation of production (and, implicit in this, the destruction of the alien relation between men and what

[1] In an outburst of voluntarism, in the Preface to the Russian edition of the Communist manifesto (1882), Marx and Engels conceded that under certain circumstances, Russia could join the socialist revolution.

[2] Lenin, *What is to be Done?* (1902). It is perhaps the most comprehensive guide-book on party voluntarism.

[3] Pareto was in touch with Sorel, and it was rumored that both had been fascinated by the promises of Mussolini, but then Pareto got disillusioned when the fascist regime nationalized big enterprises. Sorel influenced the establishment of a fascist movement in France, but later clashed with Mussolini for persecuting his old comrades.

[4] Marcuse, *From Luther to Popper* (1983, p. 149).

they themselves produce), the power of the relation of supply and demand is dissolved into nothing.[5]

Nevertheless, when it came to practice various Marxist schools supported competing approaches. They were ranging from the abolition of money circulation and all market transactions (as implemented during the initial phase of the Russian Revolution in 1917), to the planned distribution of goods (the system adopted later in Soviet Union), to allowing the limited operation of markets (adopted by some Eastern European countries and in post-1976 China).

The third, and politically most controversial, issue concerned the entity that would rein in the economy after the means of production were taken from the hands of capitalists. Different versions of Marxist theory agreed on the abolition of private property, but held widely diverging views on the institution that would replace it: would it be the workers themselves (through factory councils), the communities at large or the repulsive state? Initially, the state was condemned to extinction but ultimately won the day in actually existing socialism.

11.1.3 The Problem of "Economic Calculation"

Even before the Russian Revolution in 1917, the key question was how to design such a system and who would be the central planner to take decisions on the allocation of the means of production and on setting price levels. Expectations for solving this complex problem were running high; many believed that scientific and technological advances not only necessitated the transition to socialism, but at the same time allowed the restructuring of society on a solid scientific basis. Karl Kautsky, a famous Marxist leader, was hoping on the great help that science would provide for the construction of socialism and enthusiastically declared in 1917 that "even more than for natural science, [this is high time] for the Science of society, the so-called mental sciences".[6]

In attempting to address these issues the so-called "Russian Economic School" was eventually formed around the end of the nineteenth and early twentieth century. Despite their profound ideological adherence on Marxism, the Russian economists worked on the premise of the only available paradigm of the time: the Walrasian general equilibrium model. Discarding markets and business freedom, they vouched to solve the problem of supply and demand *a priori*, before any actual encounter between producers and consumers took place. What is sold, to whom, when and at what price, were all to be laid out in a "plan" imposed upon economic agents. Thus the economy would be in a state of perfect balance without the anathema of alienation by private property and capitalist relations.

There was, however, a caveat known in all planned economies as the problem of "*economic calculation*", i.e. how prices are set in the absence of markets? The only

[5] Marx and Engels, *The German Ideology* (1846).

[6] Kautsky, *The Materialist Conception of History* (1902).

ideologically acceptable tool to determine prices was the labor theory of value. Ricardo had established the theory and Marx had enriched it with the concepts of labor power and surplus value. The Russian economists thought that they could replace market price-setting, by determining the amount of labor incorporated in a product and then calculating the appropriate prices by a simple division to a *numeraire*. Some noteworthy exercises were undertaken by the following economists:

Vladimir Dmitriev (1868–1913) was the first to connect Ricardian and Marxist theories of value with the neoclassical theory of pricing based on marginal utility. His work was taken up by von Bortkiewicz (1868–1931) who employed the Ricardian analysis of value to calculate profit margins and relative prices.[7] Georg von Charasoff was another important economist of the time. In 1910, he laid the theoretical foundation of central planning and attempted to calculate the relationship between labor and value by using linear models of production. For obvious reasons, incorporated labor value cannot account for products which are in high demand due to fashion, rarity, etc. This led Charasoff to draw a distinction between "basic" and "luxury" products for which two different labor incorporation rates are used. All these theories, however, were based on dubious hypotheses and complicated calculations, leaving them vulnerable to many failures when tested in the real world.

Around the same period, the so-called '*cyclical theory of production*' was developed according to which the production of a commodity is the outcome of combining other commodities.[8] The new commodity is then used to produce other goods, and so on. Wassily Leontief (NLE 1973) started to design his input/output tables and was invited to work with Frisch, the founder of quantitative economic models and a keen admirer of economic planning as was examined in Sect. 10.1.[9] Based on this production cycle, economists could therefore follow the incremental incorporation of labor at each stage and calculate added value to determine the final price of the product. The difference with classic Marxist theory was that, in the cyclical model, the solution is gradual and not immediate.

11.2 Implementation of Central Planning

During the Great War, the system of central economic planning seemed not only feasible, but even natural. Many sectors of production had been commandeered for military purposes and basic goods were distributed by state authorities, not only in Russia, but in other war economies as well. Otto Neurath's proposal was the most

[7] A description of Dmitriev-Bortkiewicz model can be found in Marchionatti and Fiorini (2000).

[8] For a description see Kurz and Salvadori, *Theory of Production: A long-period analysis* (1995, Chaps. 13.2 and 13.3).

[9] At that time, von Neuman was also a student in Berlin and was working on the cyclical model. He would later flee to the US, where he developed the study of strategic decision making in the Theory of Games.

extreme: he argued not only for centrally planned production and consumption but also for the total abolition of money and the in-kind payment of workers.[10] Production should be planned using exclusively technical methods and criteria based on the use-value of goods, without any input from actual consumers.

Paradoxical as it may seem, Neurath's thought had been in line with the radical value theories of Carl Menger, the founder of the liberal Austrian School of economics. The common ground of the two polar views was that individual utility has no place in distribution, albeit for different reasons: the Austrian School dismissed marginalism as a guiding principle of individual choice, while Neurath dismissed the freedom of choice altogether.

11.2.1 Planning Failures

In the Soviet Union, economic planning passed through various stages: during an initial period (1918–1920) of the so-called "war economy". Neurath's model prevailed. Prices, money circulation, markets, commerce and individual businesses were all abolished. Consumption need not satisfy individual utility[11] but merely cover basic subsistence needs, as were arbitrarily defined by central authorities. Products were distributed directly to citizens by the state or acquired with the use of food stamps from state warehouses. It did not take long for severe food shortages to appear, followed by a famine that cost nine million lives.[12] The system finally collapsed and in 1921 was replaced by the so-called New Economic Policy which allowed some functioning of markets, though goods were sold at predetermined prices. In addressing the desperate state of the economy Lenin made an unexpected welcome to the functioning of markets:

> The chief thing the people, all the working people, want today is nothing but help in their desperate hunger and need; they want to be shown that the improvement needed by the peasants is really taking place in the form they are accustomed to. The peasant knows and is accustomed to the market and trade. We were unable to introduce direct communist distribution. We lacked the factories and their equipment for this. That being the case, we must provide the peasants with what they need through the medium of trade, *and provide it as well as the capitalist did*, otherwise the people will not tolerate such an administration.[13] [italics added]

---

[10] Neurath was the head of the Office for Central Economic Planning in the short-lived Munich Soviet Republic (April–May 1919). See Rosner (1990).

[11] Marx and Engels have a deep contempt for the concept of individual utility. As they write "pleasure-seeking philosophy has been around in Europe since the time of the Cyrenaics. This philosophy has never been anything more than the witty language of specific social groups enjoying the privilege of pleasures"; Marx and Engels, *German Ideology*.

[12] The famine was caused by a combination of prolonged draught and civil war. The Soviet Union experienced another famine in 1932–1933 due to shortages and mismanagement of arable lands. Six million people died. The source of the data is O'Grada, *Famine: A Short History* (2009, Table 1.1, p. 23).

[13] Lenin speaking in the 11th Congress of the Communist Party of the USSR. The speech is available at. https://www.marxists.org/archive/lenin/works/1922/mar/27.htm

Under central planning production and employment were substantially restructured. Workers' participation on committees that ran the factories was replaced by factory-level management, though some token participation was still maintained. Production was based on plans drawn up by a central authority which had not yet acquired the rigid structure or extensive powers that would take later.[14] This system was followed for several years, before Stalin replaced it with the 5-year plans ('Gosplan'). Initially, there were three types of plans: 15-year plans, 5-year plans and annual plans. Each plan had to meet certain interim goals within the specified period, and this soon caused many problems. Production often didn't meet the goals set out in the shorter-term plans, thus resulting in additional pressure to catch-up during the following years, whereas the longer-term ones remained unchanged so as not to undermine the unflinching faith to socialism.

11.2.2 The Gosplan

The first 5-year plan was presented in 1928 stipulating the abolishment of private land and the collectivization of agriculture. Each factory now had its own budget, recording surpluses and losses; an important innovation back then.[15] The double-entry system was also introduced, with every business activity being recorded both as an input and an output. The agency responsible for economic planning was not under the purview of the government; it was entirely controlled by the party. The new system was praised as the incarnation of Marxist theory, able to satisfy the needs of citizens and rid society of exploitation. Echoing Benthamite visions, Stalin summarized its main economic characteristics to be:

> the securing of the maximum satisfaction of the increasing material and cultural requirements of the whole society through the continuous expansion and perfection of socialist production on the basis of higher techniques. Consequently: instead of maximum profits—maximum satisfaction.[16]

Convergence with the West had been a central obsession of the communist regime, and Lenin had used a Hobbesian language to warn that "the Soviet Union cannot remain powerful, unless it becomes more powerful than the West".[17] Stalin's policy made catching up with the West a focal point and considered it as feasible, since socialist production would be more vastly more efficient and immune to the wastefulness and anarchy of the capitalist system.

---

[14] For a detailed account, see Ellman, *Socialist Planning* (1989, Chap. 2).

[15] The "ideal accounting and monitoring system" according to the words of Lenin as reported in Berend, *An Economic History of Twentieth-Century: Economic Regimes from Laissez-Faire to Globalization* (Berend 2006).

[16] Stalin, *Economic Problems of Socialism in the USSR* (1952).

[17] See Ellman (1989, *op. cit*). Subscribing to the Hobbesian approach of man's perpetual desire for power after power (See Hobbes' *Leviathan*, 1982, Chap. XI).

This led to view industrial production as the main engine of economic growth. Besides, rapid industrialization was also necessitated by war preparations and enabled the massive armaments program of the Soviet Union. As the industrial sector absorbed the lion's share of available capital, consumer goods were often in short supply, with vast numbers of citizens suffering from deprivation. Even overworking industrial workers could not expect serious material rewards—thus ideological laurels were offered instead and industrial progress was advertised as the supreme duty of society. Those managing to exceed production quotas were praised as "heroes of socialist labor"[18]; those lagging behind branded as possible dissenters. The Marxist thesis on alienation was forgotten and replaced by the regime's maxim that

> ... nowhere in the world is machinery used so willingly as in the Soviet Union, because machinery economizes labor in society and alleviates the burden of work for workers. And since in the Soviet Union there is no unemployment, workers are eager to use machinery in the people's economy.[19]

According to a central tenet in Marxist analysis, capitalism's anarchic nature implies that productive forces are often wasted. Central economic planning was supposed to do better, as the planner could use all available knowledge and information and make decisions based on science. Indeed, scientific committees were too often involved in drawing up the plans. In the beginning, the extensive use of scientific terminology and methods implied that the layman could neither decipher nor easily dismiss such an impressive jargon. The flip-side was that obscure scientific methodology disguised the cluelessness of policy makers and the failure of their projects.[20]

The most notorious case was that of Soviet pseudo-scientist Trofim Lysenko who dismissed Mendelian genetics and professed to have found a revolutionary way to improve crop yields and therefore feed more people. His experiments ruined the country's agricultural production and led to the 1948 famine, which took hundreds of thousands of lives.

In any case, the results of this massive mobilization impressed many economists in the West. The founder of development economics Evsey Domar presented in 1965 a comparative analysis of industrialization in the Soviet Union noting that

[18] Alexey Stakhanov (1906–1977) was perhaps the most famous example. He broke the mining record, exceeding the quota by 14 times, and became a celebrity and role model for all Soviet workers, appointed at the Supreme Soviet and even featured in the TIME magazine's cover in 1935. In his film Man of Marble, Polish director Andrzej Wajda shows Stakhanov as a hated figure, used by the regime as a convenient pretext to impose ever more inhuman working conditions. Later, Stakhanov's record was disputed as propaganda fabrication.

[19] Stalin (1952, *op. cit*).

[20] Centuries earlier, Thucydides had warned about the problems of elaborate planning: one is overly enthusiastic in the beginning, but at the time of execution, there is often hesitation and fear. "The confidence with which we prepare our plans is never entirely justified by their execution; speculation is safe, but, when it comes to action, fear causes failure" (*The Peloponnesian War*, A', CXVIII).

labor productivity grew in the 1928–1940 period faster than in any other country.[21] In 1950–1955 growth had slowed, falling to the same or even lower levels of western economies. The catching-up process came to a halt.

11.2.3 Chinese Planning and Failures

China was the second big state to adopt Communism in late 1940s. Its first plan was drafted by the Soviets, but from 1954 onwards the Chinese took over and focused more on agriculture rather than on industrialization. This was dictated by the urgent need to increase food production and thus put an end to widespread starvation. Nonetheless, the plan failed massively to deal with the great famine in 1958–1960 that claimed millions of lives. With China, the Soviet Union and Eastern Europe all under communism, there was a vast geographical area from the Pacific Ocean to Berlin for which central planning became the dominant model for decades to come.

The next step was to centralize central planners by imposing the Council for Mutual Economic Assistance (COMECON) throughout Eastern Europe. Although coordination with China was frequently interrupted by a growing antagonism with Soviet Union, COMECON would last until 1989, when communism collapsed and the economies entered the capitalist fold. By then, China had already adopted many features of capitalist production and embraced the operation of markets; remnants of central planning can still be found only in some Asian countries (e.g. Vietnam, Laos and North Korea).

Thus, this unprecedented mobilization of labor and bureaucratization of production, which was set to bring down capitalism, was finally swept away by the same markets that were supposed to have been eliminated. The illusion that a handful of party elites can effectively rule half the global population was replaced by the adoption of the most extreme forms of the capitalist system. In 1989–1990, on the eve of communist regimes collapsing, history had yet another twist of irony as many of the old socialist executives appeared to be ready and keen to appropriate state assets and form the new elite of private sector oligarchs. Rather than fulfilling the Marxian expectation of allying with the proletariat, they followed Pareto's elite replacement process in the most cynical way.

These developments vindicated the fears of those who had, much earlier, anticipated that the communist party would prove a hapless leader of the economy—though some of them shared the vision of a classless society. In the nineteenth century, long before the implementation of centrally planned systems, revolutionary anarchist Mikhail Bakunin had made a caustic—and prophetic—comment by wondering:

---

[21] Domar Evsey (1965), "Special Features of Industrialisation in planned economies: a comparison between the Soviet Union and the United States". Reprinted in Domar, *Capital, Socialism and Serfdom* (1989). For an evaluation of Soviet post-war successes see also in Heilbroner and Midberg, *The Making of Economic Society* (2008, Chap. 11).

...where is the mind, brilliant as it may be, or—if we speak of a collective dictatorship even if it were formed of several hundred individuals endowed with superior mentalities—where are the intellects powerful enough to embrace the infinite multiplicity and diversity of real interests, aspirations, wishes, and needs that shape the collective will of the people?.[22]

11.3 Critique of Central Planning and the Intermediate Approaches

Several economic theorists addressed the deficiencies of central planning, while it was still in operation. Some broadly supported the system and tried to make it more efficient; others capitalized on its practical problems in order to launch a polemic against Marxism. The most notable among them were the following:

11.3.1 The Criticism of the Austrian School

Followers of the Austrian School, and in particular von Misses and Hayek, had anticipated many of the shortcomings simply by their *credo* that nothing can replace the market and its price-setting mechanism. Prices convey information "signals" to economic factors which are impossible to be substituted by a committee or a computer. Naturally, the pro-market theories of the Austrian School became the rallying cry of dissenters in communist Eastern Europe; post 1989, several institutes were set up to disseminate and analyze its premises, while political parties adopted the main tenets in their economic programs.[23]

Ludwig von Mises' argument against economic planning was that capital goods can be distributed effectively only if they are privately owned. Only the owner of a finite stock of capital can calculate returns and only the market can set product prices based on their scarcity. He also argued that the central planner, by arbitrarily setting the price of one product, distorts the relative prices of other products causing misallocation of resources. His conclusion was that the state will then have no other option but to intervene and set the prices of all products, a process gradually leading to the full nationalization of the economy exactly as suggested by Neurath in 1919, during the brief soviet Republic of Munich. He wrote:

The social ideal, carried to its logical conclusion, would eventuate in a social order in which all the means of production were owned by the people as a whole. Production would be completely in the hands of the government.[24]

---

[22] Bakunin, *Anarchy: The Paris Commune and the Idea of the State* (1971).

[23] An example is the Freedom and Solidarity party (SaS) in Slovakia. Party's charter subscribes to the Austrian School of economic thought and dismisses state intervention in the markets; see Financial Times, October 9, 2011. Before, it was only in communist states that political parties unconditionally subscribed to a specific economic theory.

[24] von Mises, *Liberalism: The Classical Tradition* (2005, Chap. 2.4, "The impracticability of Socialism", pp. 46–50).

This seemingly paradoxical coincidence between staunch supporters and all-out enemies of economic planning was called "polarization on common grounds" and for years remained a hotly debated issue among economists.[25]

11.3.2 The Benevolent Planner

Another paradox is also worth noting: the Austrian School's criticism was directed not only against central planning but also against the theory of marginal utility. The representatives of the school argued that utility function is a vague and unfounded concept and is unrealistic to assume that someone goes to the market with such knowledge beforehand. The Austrian School argues that the entire problem is unsolvable as cannot even be described in objective, quantifiable terms. The inevitable conclusion is that the free market pricing mechanism alone represents the most efficient and realistic option, superior to both central planning and marginal utility theorizing.

This double assault of central planning and marginal utility makes one curious to see whether the two theories also share other common features. Marginalism assumes that consumers know the utility function and their income limitations, so it is best to let them find the balance between consumption and savings themselves. Now, assume that prices are determined by marginal utility calculation but replace ill-informed individual consumers and producers with a central planner with better information for everyone involved.

It is then easy to demonstrate that the solutions professed by economic planning are identical to the solutions rational and informed consumers would give if adopting the same utility function. However, individuals might not be fully aware of their current or future income limitations, thus some choices may be suboptimal. If there was someone with better information, such as a mighty and benevolent central planner, optimal choices for everyone involved would be made easier. What was intended by marginalists as criticism against central planning, now turns to an argument in its favor. The modern theory of consumption confirms the coincidence of decentralized and central planner solutions if perfect foresight and full information is hypothesized.[26]

11.3.3 Inventing Virtual Markets

Some economists favoring economic planning suspected the problems inherent in centrally-set pricing from early on: though the economy was relatively well

---

[25] Rosner (1990).

[26] For a mathematical proof, see Blanchard and Fischer, *Lectures on Macroeconomics* (1989, Chap. 2).

supplied with certain products, others—even of lesser value but on higher demand—had disappeared. To address this problem, they developed theories in which virtual conditions of supply and demand could determine a more appropriate set of prices and lead to a more satisfying allocation of resources.

Karl Polanyi (1886–1964), a Marxist sociologist, attempted to solve the problem of economic calculation by combining the concept of marginal utility with a system of state ownership.[27] A complex economic structure was invented in which wholesale prices where set by the government based on general production cost, while retail prices were set after negotiations between local producers and consumer associations. Producer and consumer associations composed a confederacy—a *Commune*—which regulated wages and negotiated retail margins. Realized 'profits' would then be used to finance the production of local public goods, freely distributed to its members.

Applying this system was next to impossible in the case of certain products, such as fashion or alcohol, since the price a consumer might be willing to pay could be a multiple of production cost (or zero, when an item had gone out of fashion). To deal with this problem, Polanyi adopted Charasoff's classification and conveniently labeled them as "pseudo-needs"; hence whether they were satisfied or not would have no effect on the actual wellbeing of society. Once again, the question was who would be entitled to decide which needs are 'real' and which are not, effectively back to square one of the central planning problem. Polanyi's theory was predictably attacked from two sides: supporters of economic planning argued that the confederacy of producers would eventually evolve into a central planner thus why bother with experiments. Almost the same argument was endorsed by von Misses, arguing that ultimately either the producers or the consumers would subjugate the economy to their own interests.

Polanyi's ideas were never implemented in practice, but one can find similarities between them and Oskar Lange's system, who proposed that final prices should be allowed to fluctuate in the market within a price range preset by the central planner. Lange (1904–1965) was a Polish economist and a fervent supporter of both Marxism and neoclassical theory. He attempted to solve the central planning models using an approach akin to the Walrasian process. The auctioneer is replaced by the central planner who announces the initial set of prices and then the local market, through a process of *tattonment*, sets final prices within some margin of fluctuation. This system was partly implemented by some socialist economies in Eastern Europe.

[27] Contemporary supporters of Polanyi argue that he did believe in the abolition of capitalism, but did not accept the Marxist theory premise about the transition to socialism. Gyorgy Dalos (1990) writes that, when he was in exile in Vienna, he had dismissed the soteriological aspects of socialist theory.

11.3.4 Shadow Pricing

An interesting—and technically more demanding—approach was suggested by L. V. Kantorovich (NLE 1975) in his Linear Programming framework, where a utility function is maximized under certain production constraints.[28] Employing this method, optimal production and consumption as well as the shadow prices of constraints can be simultaneously calculated. Shadow prices essentially represent the marginal utility of production resources and, by extension, what their true value would be had markets existed. This approach was, however, extremely complex as elaborate mathematical models were required for describing every single resource and commodity in the economy. As Kantorovich later admitted:

> A long-term plan... presupposes the forecast of needs, resources, [production] relations and technical capabilities... To solve these problems, further progress is needed in the model-development method.[29]

In the 1970s, central planners got some breathing space with the dissemination of computers, hoping that such complex problems could be finally solved. But hopes were dashed again as all approaches were based on hypotheses instead of real data and their results always turned out to be inferior to reality.

Even if the above moderate approaches had been implemented full scale, several insurmountable difficulties would still have remained. One important problem was that both Lange's and Kantorovich's models entailed prohibitive amounts of effort and costs to gather and process all necessary information. Another problem was that, absent financial markets, there was no mechanism of pricing investment risks, thus capital allocation decisions could not be effectively evaluated as in free markets. This resulted in planners making incorrect estimates on future returns and led, by extension, to a grave misallocation of capital. For example, it was once calculated that labor productivity in Siberia exceeded that of the workforce in the European part of the Soviet Union by 15 %. Kantorovich argued that in fact the difference was much larger—therefore more investment and resources should be channeled to Siberia. Calculations were false, he asserted, because "[p]roduct price relations in the various sectors... do not take into adequate account the different levels of income from natural resources in different areas".[30]

A third shortcoming was the so called 'Principal–Agent' problem, in which the objectives of the principal (in this case the central planner) are not dutifully followed by the agent (bureaucracy and managers), simply because their interests are different. More often than not, agents acted in their own rather than in the collective best interest, causing severe shortages, crippling planning efforts and provoking massive insubordination in factories, thus effectively bolstering the black market and their personal profiteering.

---

[28] Tjalling Koopmans (1910–1985), a Dutch mathematician, worked independently in the US on the same problem of optimal allocation of resources. Koopmans and Kantorovich were jointly awarded the NLE 1975.

[29] Kantorovich, *Essays in Optimal Planning* (1977, p. 229).

[30] Kantorovich, (*op. cit.*, p. 214).

11.3.5 The Impossibility of Efficient Pricing Under Socialism

When prices are centrally set—either according to the solutions of complex Linear Programming optimization or by defining a margin on the cost of intermediate goods—it is obvious that the market clearing price can be achieved only by chance. In most cases, there will either be an excessive supply (which would imply wasteful allocation of production factors) or an excessive demand (implying that consumers' needs are not met).

As explained by Evsey Domar, a pioneer on the theory of development, if the cost of adjusting production is high, prices in central planning tend to remain stable for prolonged periods of time, perpetuating the distortions.[31] This was also a counter-incentive for developing new products as it was simpler and cheaper to continue producing the old ones and fulfill production quotas more easily. In turn, the lack of innovation in product markets minimized the motivation to improve technology and was aggravating consumers' sense of deprivation. No matter how elaborately or brutally the regime tried to keep people uninformed, citizens were well aware of the new consumer products available in other countries, thus making the sense of dissatisfaction and disdain to multiply.

Central planning also faced additional problems from international markets, since they operated on differently-set price levels. Initially, the solution chosen was to prohibit external trade. But when trade relations with other nations resumed in 1923, the problem surfaced as how to price imported commodities. If consumers in the Soviet Union bought foreign products at international market prices, then the fearsome market daemon would have invaded from the back door and dictate prices in the socialist economy, and what then was the point of overthrowing capitalism? As the problem could not be solved theoretically, state-owned companies were tasked with handling external trade. Products were bought at international prices, but then the central planner determined their domestic price and distribution without necessarily invoking real demand. The system sometimes worked, but chronic shortages of imported spare parts and consumer goods led to bottlenecks and social disappointment.

The arbitrary pricing of imports affected budgeting at home and added further distortions and misallocation of resources. Mispricing was also common in exporting goods and raw materials (such as oil) to other COMECON economies, and both parties seemed to be content with such a practice. Soviet Union was using it to cement their dependence, while fellow governments were able to camouflage their trade deficits and presented it as a dividend of socialist cooperation.

This agreeable situation ended abruptly in 1985 when Mikhail Gorbachev, then Secretary General of the Communist Party in the Soviet Union and desperate for financial resources to stimulate domestic demand, lifted the export price discounts on raw materials. Trade deficits of countries like Hungary, Bulgaria and Poland

---

[31] Domar (1974), "On the optimal compensation of a socialist manager". Reprinted in Domar, *Capital, Socialism and Serfdom* (1989, *op. cit*).

became swollen and to finance them their governments turned to western credit. When debt levels and interest payments increased dramatically as a result of the 1987 international credit crisis, the planned economies ran into deep recession. This turn of events precipitated the collapse of socialist regimes altogether in 1989.[32] After being scornfully sidelined for years, the market mechanism prevailed full scale.

11.3.6 Sraffa's Critique

Soviet economic planning was also attacked by Piero Sraffa from an entirely different point of view. In his theory on the expanding rate of labor exploitation, Sraffa argued that the distribution of surplus in capitalism is determined by wider social conditions of production and not exclusively on the basis of workers' subsistence as originally claimed by Marx.[33] Sraffa proposed that the "double character of the wage" should be investigated by looking not only at the reproduction process but also on other factors; among them the distribution mechanism, the conditions in the labor market, the power of unions, and the motivation to remain employed in the same company.

This was an uneasy conclusion for stereotypical Marxism: not only did it suggest that under capitalism wages could further improve by enlarging their share in output, but if any of the above conditions survived under socialism it could exert exploitative characteristics on workers remuneration, no matter how triumphant the central planning appeared to be. As the ruling elite in socialist economies consisted of the political party and the nomenclature, managing the means of production could well be directed to serve their own interests as blatantly as in the system of capitalism that they were preaching to destroy.

11.4 Was Communism an Asiatic Mode of Production?

Marx's original analysis on what he called the "Asiatic Mode of Production" (AMP) would run into three types of trouble, all of them representative of the volatile relationship between party power and economic reality.

The first problem stemmed from some characteristics of AMP found in common with the Soviet model: individual property was absent and a powerful, centralized state was deciding on the production and distribution of economic outcome. These actions, however, were not sufficient to label old economies of Imperial China or Pharaonic Egypt as "socialist", given that they were relying on slavery and

[32] This is why perestroika was often called "katastroika". See Mark Almond in his article "*1989 without Gorbachev: What if Communism had not collapsed*", published in Ferguson, *Virtual History* (2011).

[33] Sraffa, *Production of Commodities by Means of Commodities* (1960, p. 9).

authoritarianism, rather than freedom and justice for all as professed by socialism—at least in the early stages of forming the theory.

To avoid such unwelcome (though truly echoing) comparisons, Stalin deleted all references to the AMP from Marxist texts. Going a step further, Georgi Plekhanov (a prominent Russian revolutionary and party intellectual) claimed that Marx himself had finally abandoned the term and did not consider it as a distinctive stage in the progress of mankind.[34]

The second problem involved another Marxist imperative that societies cannot leapfrog a stage in their historical transition but have to move one step a time: from feudalism to capitalism and only then to communism. According to this dogmatic gradualism, history is a ladder every society has to climb sooner or later before eventually reaching the socialist peak.[35] But AMP looked deviant from such a theory as it was not evolving towards either feudalism or capitalism.

New explanations were attempted to dwindle such arguments. According to Marxist historian Eric Hobsbawm, this was an oversimplified description of specific stages of social and economic development conveniently used by politicians to classify societies, but having nothing to do with Marx's original contribution.[36] Though he does acknowledge a common mechanism of transition from primitive forms of social organization to a class-based society, he argues that efforts to identify similar "general laws" in the writings of Marx and Engels were mistaken. Hence, the elimination of the "Asiatic Mode of Production" is of little consequence and, in any case, the misguided belief that Eastern economies are—from a Hegelian perspective—unhistorical and stagnant is wrong.

The above discussion reveals the impasse of stereotypical interpretations of history as a "recipe book", and the concomitant viewing of the economic process as a predetermined evolutionary path. Despite of its shortcomings, the Marxist typology of production systems does remain, nevertheless, an important contribution to the history of economic theories. John Hicks (NLE 1972), who invented the IS-LM model, acknowledging the originality of Marxist theory, wondered how it can be, that for all the progress of social sciences, almost nothing has been added to Marx's theory in a 100 years.[37]

In fact, nowadays the debate on the Asian mode of production enjoys a new life in explaining the idiosyncrasies of the Eastern economic giants: the focal point is China, a complicated amalgam of political despotism, state-owned production, absolutist labor discipline and capitalist structures—and profits. Many characteristics of the AMP are observed in other nations as well. In India, a large

[34] For an account see Godelier, *The Concept of the Asiatic Mode of Production and Marxist Models of Social Evolution* (1978).

Hobsbawm offers an alternative explanation for the Chinese Communist Party's decision to delete all reference to the AMP from official texts as wanting to avoid giving Western imperialism a reason to intervene in a "stagnant" country; see Hindes B. and P. Q. Hirst, *op. cit.*, "*Introduction*" (1979, p. 61).

[35] As Kuusinen famously put it in *Fundamentals of Marxism–Leninism* (1960).

[36] Hobsbawm "Introduction", *op. cit.* p. 61.

[37] Hicks, *A Theory of Economic History* (1969, p. 3).

part of the population remains socially isolated, while many modern capitalist businesses thrive even in regions with governments affiliated with the Communist Party. In Northern Africa and the Middle East, authoritarian regimes coexisted for many decades with decentralized structures of agricultural production and a limited number of capitalist businesses. Even today's Japan displays some characteristics of the Asian mode of production, including lifelong employment at the same corporation (often including entire families) and surprisingly uniform consumption patterns that are reminiscent of centrally-guided economies rather than capitalist-prone social attitudes.

References

Bakunin M (1971) Anarchy: the Paris commune and the idea of the state (translated and edited by Sam Dolgoff)
Berend I (2006) An economic history of twentieth-century Europe: economic regimes from Laissez-Faire to globalization. Cambridge University Press, Cambridge
Blanchard O, Fisher S (1989) Lectures on macroeconomics. MIT Press, Cambridge
Dalos G (1990) The fidelity of equals: llona Duczynska and Karl Polanyi. In: Polanyi-Levitt K (ed) The life and work of Karl Polanyi. Black Rose Books, Montreal
Domar E (1989) Capital, socialism and serfdom. Cambridge University Press, Cambridge
Ellman M (1989) Socialist planning, 2nd edn. Cambridge University Press, Cambridge
Ferguson N (2011) Virtual history. Penguin, London
Godelier M (1978) The concept of the "Asiatic mode of production" and Marxist models of social evolution. In: Seddon D (ed) Relations of production: Marxist approaches to economic anthropology. Frank Cass, London, pp 209–257
Heilbroner R, Milberg W (2008) The making of economic society. Pearson Education Inc., Englewood Cliffs, NJ
Hicks J (1969) A theory of economic history. Oxford University Press, Oxford
Hindes B, Hirst PQ (1979) Pre-capitalist modes of production. Routledge & Kegan Paul, London
Hobbes T (1982) Leviathan. Penguin, New York
Kantorovich LV (1977) Essays in optimal planning. Blackwell, Oxford
Kautsky K (1902) The aims and limitations of the materialist conception of history. Available at https://www.marxists.org/archive/kautsky/1902/08/aims-limitations.htm
Kurz H, Salvadori N (1995) Theory of production: a long-period analysis. Cambridge University Press, Cambridge
Kuusinen OW (1960) Fundamentals of Marxism – Leninism. Foreign Language Publishing House, Moscow
Lenin VI (1902) What is to be done? Foreign Languages Publishing House, Moscow
Marchionatti R, Fiorini R (2000) Between Walras and Ricardo: Ladislaus von Bortkievicz and the origin of neo-ricardian theory. Revue européenne des sciences sociales. Tome 38 (117):173–191
Marcuse H (1983) From Luther to Popper. Verso, London
Marx K, Engels F (1846) The German ideology. Progress Publishers, Moscow
O'Grada C (2009) Famine: a short history. Princeton University Press, Princeton
Rosner P (1990) Karl Polanyi on social accounting. In: Polanyi-Levitt K (ed) The life and work of Karl Polanyi. Black Rose Books, Montreal
Sraffa P (1960) Production of commodities by means of commodities: Prelude to a critique of economic theory. Cambridge University Press, Cambridge
Stalin J (1952) Economic problems of socialism in the USSR. International Publishers, New York
von Mises L (2005) Liberalism: the classical tradition. Liberty Fund, Indianapolis

From Keynesian Economics to Stagflation 12

Abstract

Marxist economists were not alone in their complacency that cycles are ruled out under central planning. The stability of post-war economies has made Governments and mainstream economists alike to believe that Keynesian fine-tuning is able to thwart any sort of a recessionary threat. In a parallel development, microeconomics was dealing with all problems of individual behavior and market imperfections so that distortions could be minimized and a high level of prosperity would be constantly attained. The disintegration of the Bretton Woods system opened the Pandora's Box of economic malfunctions. Soon, Western economies were engulfed in the oil-price turbulence and a novel type of crisis that pushed both unemployment and inflation upwards. This development shattered dominant theories and ushered in the era of monetarism, rational expectations and strategic behavior. But as unemployment was piling up, economic theories and governments had to pay more attention to market imperfections and this enabled a more consensual approach among the old rivals of Keynesianism and Monetarism.

12.1 The Postwar Golden Years

Post WW II challenges were immense as national economies in Europe, Japan, Russia and China had ended in ruins. German factories had been destroyed by Allied bombing, while British cities had been devastated by German air strikes. A colossal restructuring and reconstruction effort was needed in Europe.

12.1.1 Reconstruction and Stability

The United States, not having suffered any domestic damages during the war, headed the rebuilding program. Reconstruction took place in an environment of

N. Christodoulakis, *How Crises Shaped Economic Ideas and Policies*,
DOI 10.1007/978-3-319-16871-5_12

remarkable stability in currency markets after the Breton Woods agreement was signed in 1944. Signatory governments agreed to adopt a monetary policy that would maintain (almost) stable exchange rates by tying their currencies to the US Dollar, which was convertible to gold. The US dollar, once more, had become the most powerful currency in the world. *In tandem*, the US emerged as a superpower in the new global order. Western Europe would also embark on a long period of economic stability and growth, especially after a major currency reform was implemented in West Germany in 1952.

As their countries were mired in poverty, southern Europeans migrated *en masse* to the United States and Northern Europe. In 1946, Greece became embroiled in a civil war; Italy's reconstruction focused on the North, leaving the South poor and at the mercy of the Mafia,[1] while Spain and Portugal were ruled by prewar dictatorships. In an attempt to alleviate war suffering, postwar European governments provided grants and public services to their citizens, paving the way for the emergence of the welfare state; in England, the National Health System established in 1948 became the hallmark of the new social paradigm. As universal suffrage had by then become the norm, welfare recipients acquired a voice to express their satisfaction or discontent. Thus governments were motivated to protect the weakest groups and stimulate the economy not just for advancing the Benthamite common good, but also on the expectation of being rewarded at the next election. The new economic theories analyzed and evaluated policies focusing on the way the economic product was enhanced and distributed more fairly among social groups.

Post-war reconstruction efforts allowed the West to experience a long period of economic stability and sustained growth; unemployment and inflation remained low and the welfare state was constantly expanding, marking the period 1950–1980 as the "*Golden Era*" of Western history, according to Hobsbawm.[2] Economic theories naturally focused on how to preserve stability and broaden the path to prosperity.

12.1.2 New Divisions

Elsewhere, radical changes were also taking place. Many former colonies gained independence, undercutting the imperialist potency of England, France and Belgium, and later that of Portugal. Global exchange relations also underwent vast changes, with several poor countries seeing trade as a grand opportunity for their economic development.

---

[1] Mussolini had hunted down the Mafia in the interwar period; see Smith, *Mussolini* (1982, p. 93). After the war, however, the allies used its network for food aid distribution, since they did not trust the state mechanism, and the Mafia was made powerful again.

[2] Hobsbawm, The Age of Extremes: *the short twentieth century, 1914–1991* (1999, Chap. 9).

Not all developments, however, went in the direction of universal progress and soon new divisions sprang out, domestically, regionally and worldwide. Moreover, the whole world was divided into two power blocs, the Eastern one dominated by the Soviet Union and the Western one dominated by the United States. Because of the sustained rivalry and tension between the two blocs over the next 40 years, named as "the Cold War" period, a very large amount of resources was allocated to manufacturing of new armaments which put additional pressure on reconstruction.

The super-power antagonism led to some heavy-handed interventions in third countries. In 1953, the democratically elected government of Mohammad Mosaddegh wanted to nationalize the British-controlled oil industry in Iran, a move that produced shock waves in the West worrying that the oil nationalization campaign might culminate in a communist transformation of the whole area. A quickly organized coup overthrew the government and oil prices stayed at very low levels for two decades, facilitating the rapid economic growth of Western economies. But everything comes at a price: the events in Iran quickly backfired and a wave of national emancipation movements followed.

A new rival to western interests would emerge with the Suez Crisis of 1956, when Egypt's President Nasser decided to nationalize the Canal, a vital trade route for the West. Britain and France responded by assuming military action, but finally they were forced to withdraw. Their defeat marked the weakening of European influence in the global arena, to the extent that prompted then Chancellor Adenauer of Germany to declare that only a united Europe could have a future. Two years later, in 1958, six European nations (Germany, France, Italy, the Netherlands, Belgium and Luxembourg) established the European Economic Community, which proved to be a milestone in restructuring Europe and safeguarding peace in the continent.

12.2 The Mixed Economy and New Theories

Sea changes also occurred in the ownership structure of the postwar economy. Until then big enterprises were privately owned, except of course the nationalizations following the socialist revolution of 1917 in Russia. But the war changed the options afterwards: major industries were nationalized in Britain, including the railways, ports, the power company and the Post Office. At that time, state-control meant that a lot more resources as compared to private initiatives could be mobilized for the reconstruction effort and, thus, public utilities expanded at an unprecedented rate. The same model was followed in most Western economies, though not in the US where corporations by and large remained privately owned.

12.2.1 The Division of Economics

Depending on the public ownership status, three types of economies emerged in the postwar landscape: centrally planned economies with full state ownership as examined in the previous chapter, free-market economies with a dominant private sector, and mixed economies with state-run enterprises coexisting with the private sector.

Markets underwent a similar transformation. Until then, economic theory was premised on the absolute freedom of markets, but after the war alternative models emerged. A broader discussion concerning markets and pricing was underway, focusing on how to deal with oligopolistic practices and alleviate their distorting consequences on consumers.

In the postwar ventures of economic theories, a focal point was the formulation and implementation of policy: how specifically can governments use the tools available to them to maximize employment and keep growth uninhibited? These questions were addressed by transforming Keynesianism from a crisis-recipe in times of contraction to a panacea of all-time state intervention aiming at full employment by boosting demand and domestic production.

The pursuit of those two goals divided economics into two separate fields of study: one was *micro-economics*, tasked with analyzing individual behaviors, market structures, and corporate ownership; *macro-economics* on the other hand dealt with aggregate economic phenomena, including unemployment, inflation, crisis dynamics, etc. This dichotomy still remains to date, though the divide is no longer as deep.

The various combinations of state-ownership and market regulation levels meant there was by now a great diversity of economic systems and theories. The US economy before 1929 combined private business ownership with free markets. Post-1929, the ownership model remained more or less intact, but markets came under extensive regulation and intervention by the Government. Pre-1940, the British economy followed a model of free markets and privately-run enterprises. During the War, it adopted several planned economy features to fight hoarding and black market practices, while after the War many enterprises were nationalized and their respective markets regulated. The shift from one pattern to another and the need to assess their outcomes motivated the emergence of comparative and transition economics as new fields.

Economics could finally boast to be a most prestigious science: it had a rich theoretical framework, powerful analytical tools, sophisticated forecasting methods and optimal solutions suited to any taste of market and state combinations. The crash landing occurred when a new unknown phenomenon, code-named *stagflation*, caught everyone by surprise in the 1970s.

12.2.2 The End of Postwar Stability

When US financial resources became overstretched by the Vietnam War, the Breton Woods agreement became problematic. After a typical monetary expansion to finance the cost of the war, inflationary pressures piled up on the dollar. With other currencies remaining pegged to the US dollar, the latter's convertibility to gold led many holders to frantically interchange them. Gold prices spiked increasing the pressure on the sustainability of the system. In 1971, the United States could no longer afford the depletion of reserves, suspended convertibility and devalued the dollar by around 10 %. To sustain the new exchange rates, monetary authorities organized major interventions in the currency markets, but these proved to be inadequate and in March 1973 the Breton Woods system finally collapsed.[3] Every country was free to devalue its currency, as it had done in the 1930s.

Worried by the volatility of their dollar denominated revenues, the oil producing nations established OPEC, the Organization of Petroleum Exporting Countries, and agreed to sell at a common price worldwide. Cartel practices sent the price of oil skyrocketing, dramatically increasing the inflation pressure already caused by serial currency devaluations in the post-Breton Woods era. Interestingly enough, the sudden monetary chaos did not evolve into full-scale economic crisis as in the 1930s. This was avoided for two reasons:

First of all, the dollar was weaker, but it had not lost its status as a world reserve currency. There were other hard currencies, such as the German mark or the Swiss franc, but their influence on international markets was limited and neither country had the aspiration to assume the role of reserve supplier worldwide. The second reason concerned growth: the world economy was riding high at the time, thanks mainly to the expansion of international trade, which meant that competitive currency devaluations were not as catastrophic for their neighbor's exports.[4] As can be seen in Fig. 1.1, the US underwent a brief period of contraction and soon growth had returned. Monetary cooperation was also a standard feature in the European agenda, leading to innovative schemes of national currency coordination, like the "snake in the tunnel", a predecessor of today's single currency.

The oil and monetary developments in the 1970s signaled the emergence of new economic forces. Oil-producing countries accumulated massive revenues and started becoming major financial actors worldwide. The most remarkable economic transformation took place in China. After Mao Zedong died in 1976, Deng Xiaoping assumed party leadership two years later and adopted the so-called "four modernizations" (in industry, agriculture, science and technology and national defense), gradually bringing China into the fold of international economic system. Key in the reform process was the introduction of a western-type system of national accounting and GDP measuring. As was the case with Meade and Stone in Britain in the 1930s and with Kuznets, Schulz and Mitchell in the United States in the 1920s, the new system was designed by economist Fengbo Zhang provided China with a coherent framework to seek stability, growth and credibility.

Other parts of the world were less fortunate in jump-starting their economies. Several countries in Africa and Latin America underwent periods of great economic instability, typically followed by authoritarian politics and social repression.

---

[3] Garber (1993) claims that the collapse was one of the most accurately and widely predicted events in economic history and cites several warnings by economists.

[4] For a more detailed and interesting account of this short-lived recession, see Eichengreen, *When Currencies Collapse* (2012).

Several economic theories purporting to defend or attack the new realities came in the front.

12.3 The Stagflation Crisis and the Monetarist Counterattack

In the field of economic theory, a spectacular overturn took place after the combined effect of stagnation and high inflation (called '*stagflation*') discredited the dominant Keynesian paradigm. Governments initially attempted to address the economic slowdown by using policies that in the past had been miraculously effective: a combination of fiscal expansion to bolster demand and monetary easing to bring down interest rates. With unflinching faith to the Philips curve tradeoffs between unemployment and inflation, they expected that inflation would probably rise a little, but at least unemployment would drop and price hikes could later be arrested by some monetary tightening. Hitherto, the *credo* was that stagnation goes hand in hand only with deflation (i.e. a very low or even negative inflation rate caused by depressed demand), so when both skyrocketed the recipe turned sour.

Enter a new school of economic thought—*monetarism*. It was developed and disseminated by economists mainly at the University of Chicago, and soon became immensely popular among academics and policy makers in the US and Europe. A three-front attack against Keynesian orthodoxy was launched along the following issues:

(a) *Type of disruptions*: According to the monetarist approach, divergence from full employment was not caused by insufficient demand but by supply-side shocks. Therefore, Keynesian stimulus policies are ineffective and only lead to higher inflation, deficits and debt levels. This analysis paved the way for the Real Business Cycle theory, subsequently developed by Kydland and Prescott (NLE 1998). According to the new theory, fluctuations in economic activity and other economic fundamentals are accounted for by external disruptions, technological advances and changes in consumer preferences. Since none of these problems can be effectively dealt with by state intervention, the economy should be left to return to a state of equilibrium simply by the adjustment in wages and investment activity. Otherwise, state intervention can make matters worse, by postponing the necessary adjustment of production factors.
(b) *Monetary policy*: When the economy slows down due to a supply-side disruption, monetary easing leads on a par to rising inflation, as the quantity theory of money stipulates. A totally different policy recipe was suggested by tightening money supply to fight inflation. Only then, theory predicts, confidence can return and the economy resumes growth.
(c) *The Phillips curve*: Until then, the relation between inflation and unemployment rates was explained by the celebrated Phillips curve. In the traditional model, economic agents (for example firms and/or unions) are unable to perceive the interactions of the system and make inaccurate inflation projections in setting prices or wages respectively. But as prices rise in the

aftermath of a demand-push policy, real wages drop and it is precisely this mechanism that temporarily enhances economic activity.

Robert Lucas (NLE 1995) rejected these assumptions as false and contested that if economic agents are "rational" they can anticipate the strategy of the government and demand a higher pay rise. This eliminates the element of surprise, the Phillips curve now becomes vertical and unemployment remains at the same high level. This is the so-called non-accelerating inflation rate of unemployment, (NAIRU). This realization fell like a bomb and rattled Keynesianism for a long time, before a more realistic synthesis between the Keynesians and the Chicago economists emerged later.

After the supply-side crisis of the 1970s, economic policy was never again a unilateral decision by the Governments but had to take into account expectations, rival strategies and a number of constraints, both domestic and external ones. New theoretical advancements adopted a more strategic role for economic agents and led to the design of new institutions to spearhead stability and credibility. The neoclassical synthesis provided policy makers with the best of the two worlds, safeguarding monetary stability and providing some fiscal room for facing mild recessions.

12.4 Theories of Strategic Behavior

The idea of competitive strategies was originally developed by von Neumann[5] who had abandoned his interwar aspirations in Berlin over central planning uniformity and formulated the so called "game theory" where economic agents are assumed to compete with each other. Later John Nash (NLE 1994) developed several formal models of conflict and competition, giving a new impetus and applicability to the theory. Eventually, the theory of games provided a framework of analyzing economic interactions in terms of strategic interaction between the actors, and not any more as an entirely deterministic system like those found in physics where one applies an input, observes the results and intervenes again until the desirable outcome is achieved.[6]

Lucas's critique (1976) demonstrated that if economic agents are not passive, they can act preemptively and frustrate government's strategy. In the vacuum created by the retreat of Keynesianism, a new macroeconomic theory emerged, which dismissed passivity and analyzed economic systems on the basis of interaction and strategic competition, i.e. by efficiently forming expectations. The way expectations are formed reveals a lot about the kind of strategy employed by economic agents. When expectations are derived retrospectively or by taking official announcements for granted, economic agents in effect surrender the

[5] von Neumann and Morgenstern, *Theory of Games and Economic Behaviour* (1944).

[6] This is the so called feedback mechanism, which gradually leads the deterministic system to balance, so long as certain stability conditions are met.

initiative to some other entity which makes its own decisions by taking the other's response as given.

An example of such leader-follower situation is the traditional interpretation of the Phillips curve. The trade-off between unemployment and inflation is explained by the ability of the leader (in this case, the government) to boost demand before the followers (in this case, the unions) are in a position to adjust demands and negotiate a higher nominal wage. It is only with misguided economic agents that real labor costs may become lower and lead to production increases. But if, instead of thinking retrospectively, economic agents anticipate the strategy of monetary authorities the Phillips Curve is incapacitated: expecting higher inflation, unions negotiate higher wage increases from the beginning, thereby negating the lower costs businesses were expecting after the monetary expansion.

12.4.1 Strategy and Credibility

Game theory applications are found today in several fields of economic theory, including industrial organization, the theory of public choice, theory of contracts, the principal-agent problem, etc. It is also used in the analysis of social behaviors, political action, group negotiations, and many others.[7] In game theory, the interactions between economic agents can be examined under the perspective of either competition or cooperation. The outcomes for the parties involved are usually worse when they are competitive, but still competition dominates because the other part is not trusted to be cooperative. When both sides are suspicious of each other, both end up being worse off. Economists were then challenged to explain how agents can be elicited to act cooperatively instead of antagonizing to their mutual detriment.

Credibility and confidence play an important role in the case of economic crises. Examining this effect, George Akerlof and Shiller (NLE 2001 and 2013 respectively) coined the term "*confidence multiplier*": falling confidence results in more and more households spending less, ultimately leading to a general collapse of demand.[8] Bank runs also happen when there is little confidence among households in the banking system's viability. The panic of massive withdrawals after the 1929 crisis was fueled by the memory of bank defaults in previous, less severe crises.

12.4.2 Institutions and Crises

The quest for cooperative outcomes and credibility led to the coining of new concepts in economics, such as commitment technology, reneging, retribution

[7] For a collection of insightful papers on game theory applications see the volume by Diekmann and Mitter, *Paradoxical Effects of Social Behavior* (1986).

[8] Akerlof and Shiller, *Animal Spirits* (2009).

and signaling, and the establishment of institutions tasked with imposing compliance with the cooperative framework and minimizing the incentive to break the agreement. Customs unions, for example, prevent trade wars between participating countries. In the case of monetary unions, member states cannot implement unilateral devaluations of their currency while, by ceding monetary policy to an independent Central Bank, governments no longer have access to the printing press, and so inflation-fighting policies can become credible.

Models of rational choice can also explain how a crisis is sparked. Game theory predicts that situations with only a few players where processing information is easier reach more efficient outcomes in comparison with a large number of agents where things become complicated and inefficient.[9] In the event of a crisis, mass markets may turn irrational because information is difficult to obtain and evaluate. With confusion rising, social trust is thinning out and crowd behavior follows. Crises can happen even when credibility institutions exist: if it is uncertain whether they survive and agents are, therefore, not sure if other players will stick to the rules or rush to break them.

If an institution does collapse, those who rush to exit will suffer the smallest loss (or reap the largest benefit), as in the case of a fire when people are in danger of also being trampled upon. In this case, the crisis can be the result of everybody rushing to the exit, in a self-defeating behavior. Examples include the unruly collapse of the Gold Standard in the 1930s and the panic-stricken dismantling of the EU's exchange rate mechanism in 1992–1993. Two conflicting approaches were developed to deal with the danger of a crisis caused by a systemic collapse:

(i) One theory argues in favor of abolishing all constraining rules and letting any response free to happen. In a free-floating currency regime, for example, countries have nothing to fear of a non-existing mechanism that would restrain movements in the exchange rate.
(ii) The other approach argues in favor of tougher rules and obligations that effectively take the choice of leaving the institution off the table. The deeper a union is formed, the more it is ensured that member-states respect the rules of the game and do not consider escape strategies. The euro is a good—though not yet fully accomplished—example: members have much more to lose from an exit that they would gain from recovering their monetary independence.

Numerous intermediate approaches have also been developed between the two ends of all-free and too rigid commitments respectively. These approaches combine a set of rules with some degree of flexibility. The Breton Woods system and target zones of exchange rate were examples of such compromises.

[9] See Shubik, *Game Theory in the Social Science: Concepts and Solutions* (1985, Chap. 12).

12.4.3 Rational Choice in Social Sciences

Rational choice theory proved to be a very useful tool in analyzing various forms of social group behavior, from class competition to open conflicts. In relation to economic policy, Mancur Olson employed the rational choice framework to argue that individuals tend to form pressure groups to promote their special interests. This is to the detriment of public goods, as they do not allow the economy to take advantage of new technologies, open commerce and competition. He explained this by arguing that pressure groups reap the benefits of the status quo today and will suffer a loss if things change, while those who stand to benefit have no voice to influence policymakers. Olson's analysis heavily influenced collective bargaining theories, according to which reforms are facilitated when the affected parties are reasonably compensated.[10]

Perhaps the most controversial use of rational choice theory was conducted by Gary Becker (NLE 1992) who extended the theory to analyze social issues and patterns such as marriage, inheritance, divorce, and crime rates. His central argument was that behaviors can only be understood if preferences and available means are taken into consideration in a rational calculation. When Becker proposed the legalization of trade in human organs as a means to eradicate black market and ensure that poor donors are at least paid instead of being murdered, he was attacked for serving the rich who can afford to pay for the organs, disregarding any sense of social solidarity. But in other issues, Becker's insights were more agreeable. On the low fertility rates observed in developed countries, he explained that higher real wages and career considerations increased the opportunity cost of parenthood for adults, and especially so for women.[11] Fertility rates rise in times of economic contraction, because the cost of spare time drops.

The originality of his approach left a deep impression in economic theory though it was frequently attacked or even mocked. For example, Amartya Shen[12] ridiculed his theory by devising the effigy of a "rational fool": when a person is asked by a stranger for directions to the railway station, the 'fool' maximizes individual wellbeing by pointing to the Post Office and requesting that a package be dropped off as a favor. The stranger pretends to accept, with the intention to steal the parcel immediately after!

[10] Olson, *The Logic of Collective Action: Public Goods and the Theory of Groups* (1965, Chap. V).

[11] Disproving Malthusian theories. See Becker, *Human Capital: A Theoretical and Empirical Analysis, with Special Reference to Education* (1993, Chap. II, p. 23).

[12] Sen, *Rational Fools* (1979).

12.5 The Deregulation Crisis and the Keynesian Counterattack

The conflict between Keynesianism and monetarism had dramatic political ramifications in the 1980s, as the governments in the UK (under Margaret Thatcher) and the US (under Ronald Reagan) were elected on the prospect of endorsing the doctrines of the Chicago School. Both governments embraced monetary tightening policies, which were successful in fighting inflation but at the cost of inflicting much higher unemployment. The same governments also set out to dismantle the welfare state, fully privatize public enterprises, abolishing collective bargaining and liberalizing the labor market. More generally, their aim was to minimize government intervention and deregulate markets. It did not take long for the weaknesses of this policy to appear: unemployment remained high for long periods, income inequalities increased and growth failed to take-off on a sustained path.

12.5.1 The Neoclassical Synthesis

In the 1990s, the pendulum of economic policy seemed to be distant from both the neoliberal faith on minimal government and Keynesian overconfidence in the potency of a large one. The result was the so-called "neoclassical synthesis", which is currently the prevailing theory in macroeconomics. In its long term analysis, this synthetic framework takes its cue from the neoclassical model in which economic fundamentals change according to factor returns and monetary policy eventually becomes ineffective. In the short term, the synthesis is broadly Keynesian, since prices are rigid and monetary policy can briefly influence demand levels and jump-start economic activity.

This allowed the so-called micro-foundation of macro-paradigms, adding theories of individual behavior to aggregate economic models. With this compromise, monetary policy can be effective until contracts adjust to the new conditions. If policymakers take advantage of this window of opportunity and actively support private investment (with infrastructure projects, technology dissemination policies, etc.), this short-term boost of economic activity can acquire more permanent characteristics and lead the economy out of recession. A well-known offspring of the above synthesis is the way monetary policy is conducted today by several central banks. According to the so called Taylor's rule, interest rates are set at a level that serves inflation targeting (the monetarist component) and helps economic activity as well (the Keynesian component).[13]

[13] The rule in its current form was proposed by Taylor, *Discretion versus Policy Rules in Practice* (1993).

12.5.2 Market Freedom and Equilibrium

The debate on markets never ceased. The Austrian School defended the uncompromised freedom of markets as the only way to safeguard the free choice of individuals. Any attempt of market intervention claiming to serve the interests of the majority is rejected as dangerous moral subjectivism. Echoing Hobbes who had called Aristotelian teaching a "vain philosophy", Hayek branded any such attempt to be a "vain conceit" that results in having an unaccountable bureaucracy assuming control and enslaving individuals.[14] The Austrian School's position was strengthened by General Equilibrium theory, according to which only free markets can ensure the optimal allocation of resources in the sense described by Pareto.

Some issues, however, had remained unanswered: first, the Walrasian equilibrium model was based on the *tatonnement* mechanism with no mathematical proof as yet, and, second, the uniqueness of such an equilibrium was not guaranteed either. A third problem relates to the issue of stability of equilibrium, and, finally, the point of equilibrium must be calculable so as to be utilizable. All four issues were addressed by Kenneth Arrow and Gerard Debreu (NLE 1972 and 1983 respectively) by proving two remarkable theorems:

- The first theorem showed that if a general equilibrium does exist, then there can be no other point where one's gain does not imply another's loss. In other words, the general equilibrium is the unique solution that meets Pareto's criterion.
- The second theorem showed that the equilibrium state is stable, in the sense that if disturbed it returns to the initial position. However, the assumptions are so delicate that the system could easily produce non-converging fluctuations around equilibrium.[15]

Under another set of assumptions, the general equilibrium is also calculable—through a complex calculation, but nevertheless a feasible one. These developments were considered a great revolution in economics, since they seemed to be solving all the problems of free competition and social wellbeing in one stroke.

These theorems, however, depend upon many assumptions, not all of which look plausible in the real world as, for example, the prevalence of perfect competition and perfect information. But even under unrealistic assumptions, it is perhaps useful to know the equilibrium state of an economy and use it as a benchmark to calculate how much consumers lose when market competition is not perfect. It also allows policymakers to assess whether a market needs to be reformed or not. In this context, a market failure implies that distortions may waste opportunities, thus an alternative arrangement and allocation of the same resources can bring about a better outcome for everyone involved.

[14] This theory is elaborated in Hayek's famous book, *The Road to Serfdom* (1944).

[15] See for example Scarf (1960).

12.5.3 Market Socialism: An Attempt of Synthesis

One response is to see market distortions as a failure of coordination between sectors of the economy. Under this approach, economic agents read the "signals" differently from ongoing changes in supply, demand and pricing. The result is that not everyone is able to adapt promptly and sufficiently, since agents have variable reactions and adaptation capabilities. This realization gave birth to a new strand of economic theories in the late twentieth century, dealing with market failures and the appropriate policy responses. State intervention can take two forms: monitoring and regulating the markets, and directly stimulating supply and/or demand to compensate for the deficiencies.

In other cases, inferior market outcomes stem from social deprivation caused by low incomes and the lack of ownership of capital goods—in simple words, it is the result of pre-existing poverty. In this case, individuals are unequal market participants, as a result of the original backlash. Many theories were developed to alleviate social inequalities, proposing measures such as the imposition of estate taxes to redistribute wealth from privileged to underprivileged citizens; provision of income support to poor households through the welfare state; provision of poll grants to ensure a basic standard of living for all (for example through the establishment of minimum wages and pensions).[16]

Even when it is possible to alleviate the original backwardness, the market mechanism is likely to create new inequalities in the production and distribution stages. To deal with the former, the provision of education and training to the workforce allows workers to improve their skills and utilize opportunities in the job market. Workers' participation in company's profits improves deficiencies in distribution, as otherwise production benefits brought about by new technologies, cheaper raw materials or increased demand are usually reflected on profits rather than wages. In account for this, a framework of collective bargaining was devised by including agreements that give all employees a share of company's net profits, echoing the approach Mill had made in the nineteenth century.[17]

But this well-indented approach quickly becomes problematic: during an economic boom, workers resist new hires so as to keep higher earnings for themselves and this causes the company to become stagnant and rigid, unable to take advantage of new opportunities. During a contraction, workers fight for their jobs which means that either their income will shrink or the entire company goes bust. In an attempt to deal with these problems, James Meade developed a theory of labor-capital partnership, according to which participants agree how to allocate the risk of income fluctuations instead of bargaining the wage level. Businesses are run by boards consisting of equal numbers of owners and workers. Capital is divided into owners' shares, which enter the stock market, and workers' shares, which are not traded.[18]

---

[16] For a detailed account, see Le Grand and Estrin, *Market Socialism* (1989).

[17] For details, see Weitzman *The Share Economy* (1984).

[18] Meade, *Alternative Systems of Business Organization and of Worker's Remuneration* (1986).

This idea is similar to the bonuses offered in times of high profitability and the labor-capital negotiations in times of crisis, when workers agree to lower wages in exchange for keeping their jobs. James Meade later suggested an economy-wide system of decentralized bargaining, under which wages are set after considering the unemployment rate at a local and general level, as is the case with decentralized collective bargaining systems.[19]

References

Akerlof G, Robert S (2009) Animal spirits. Princeton University Press, Princeton

Becker G (1993) Human capital: a theoretical and empirical analysis, with special reference to education. The University of Chicago Press, Chicago

Diekmann A, Mitter P (1986) Paradoxical effects of social behavior: essays in honor of Anatol Rapoport. Physica-Verlag, Berlin

Eichengreen B (2012) When currencies collapse. Foreign Aff 91(1):116–134

Garber P (1993) The collapse of the Bretton Woods fixed exchange rate system. In: Bordo M, Eichengreen B (eds) A retrospective on the Bretton Woods system: lessons for international monetary reform. University of Chicago Press, Chicago, pp 461–494. http://www.nber.org/chapters/c6876

Hayek F (1944) The Road to Serfdom. Routledge, London

Hobsbawm E (1999) The age of extremes: the short twentieth century, 1914–1991. Abacus, London

Layard R, Nickel S, Jackman R (1991) Unemployment: macroeconomic performance and the labour market. Oxford University Press, Oxford

Le Grand J, Estrin S (1989) Market socialism. In: Le Grand J, Estrin S (eds) Market socialism. Clarendon, Oxford

Meade J (1986) Alternative systems of business organization and of worker's renumeration. Allen & Unwin, London

Olson M (1965) The logic of collective action: public goods and the theory of groups. Harvard University Press, Cambridge

Scarf H (1960) Some examples of global instability of the competitive equilibrium. Int Econ Rev 1 (3):157–172

Sen A (1979) Rational fools. In: Hahn F, Hollis M (eds) Philosophy and economic theory. Oxford University Press, Oxford

Shubik M (1985) Game theory in the social sciences: concepts and solutions. MIT Press, Cambridge, MA

Smith DM (1982) Mussolini. Knopf, New York

Taylor J (1993) Discretion versus policy rules in practice. Carn-Roch Conf Ser Public Policy 39:195–214

von Neumann J, Morgenstern O (1944) Theory of games and economic behavior. Princeton University Press, Princeton

Weitzman M (1984) The share economy: conquering stagflation. Harvard University Press, Cambridge, MA

[19] Layard and Nickel, *Unemployment: Macroeconomic Performance and the Labour Market* (1991, Chap. 2, Wage Bargaining and Unions).

Development, Collapse and New Theories 13

Abstract

Economic growth became the catch word for many countries in the postwar period, no matter their ideological preferences and political realities. New theories try to explain the laws of capital accumulation and how the process can be improved so as to ensure a balanced and sustainable trajectory. Two off-springs of Keynesianism offered competing explanations on capital accumulation, while countries in Latin America attempted to achieve it by curtailing imports and mobilizing domestic resources. In most cases, inward economic development was mired with rising indebtedness and countries ended up with heavy-handed adjustment programs and external surveillance. In other cases with a thin domestic accumulation, development was encouraged by international aid. Yet again, misallocation and abusing of foreign resources trapped many poor countries into high debt. Economic theories faced several challenges on how a debt-crisis is efficiently handled and what kind of institutions can minimize the recurrence of malpractices.

13.1 Theories of Economic Development

The question of development became urgent for many countries during the interwar period and economists had to figure out how, not only escape the recessionary waves but also, follow a path of sustained growth without the kind of economic instability witnessed after the crash in 1929.

13.1.1 The Warranted Growth Model

Working independently, Roy Harrod in 1939 and Evsey Domar in 1946 developed the first analytical model of economic growth in terms of savings levels and capital productivity. By assuming that marginal capital productivity remains stable, they

N. Christodoulakis, *How Crises Shaped Economic Ideas and Policies*,
DOI 10.1007/978-3-319-16871-5_13

made the distinction between the "warranted" rate of economic growth versus the "natural" rate of population growth. If the two rates are equal, the economy is in a state of full employment growing as fast as the labor force.

However, when the warranted rate of growth is higher due, for example, to excessive savings and investments the result is overproduction and a Keynesian-type of crisis is ensuing. When savings shortages stifle investment, the result is excessive demand and insufficient supply. The appropriate policy response is to increase capital investment and make savings more attractive. The Harrod–Domar model heavily influenced economic thought and eventually led to several policy suggestions to spurn growth. For example, if some nations were so poor that savings were insufficient to fund investment they had to find an alternative way to accumulate capital, either through foreign lending or from international aid.

13.1.2 The Solow Growth Model

The next theoretical landmark was the growth model developed by Robert Solow (NLE 1987) and Trevor Swan. In this case, economic growth is measured in terms of new investment, the creation of new jobs and advancement in technology. In contrast to the Harrod–Domar framework, the marginal efficiency of capital in the Solow–Swan model is diminishing eventually leading to the convergence between developed and developing economies without external intervention. According to the theory of diminishing returns, nations with a substantial level of initial capital had undergone a period of rapid economic growth but then growth rates gradually declined. On the other hand, nations with insufficient initial capital accumulation offer higher capital returns resulting in faster growth of their economies as compared with more advanced countries.

Such was the great expectation of the post-war era: if poor countries succeeded in accelerating their growth rate more than rich ones, the gap separating the two worlds would gradually narrow. But reality often disproved this certainty, forcing economists to develop alternative models to explain the deviations and suggest alternatives: one such effort was the "threshold theory" according to which a minimum level of initial capital accumulation is necessary to kickstart growth. Since poor countries do not have this initial capital, the international community should help them by providing financial aid.

This realization led to the establishment of the World Bank after the end of World War II, an institution tasked with providing loans to poor countries helping them to accumulate the necessary capital. Other competent institutions, including the United Nations and a host of smaller agencies were also established with the same mission. These initiatives proved successful in several cases with war-torn economies starting to grow very quickly; one of the most successful examples was the Marshall Plan, an American aid program that helped rebuild European economies after the war.[1] In other cases, however, aid programs failed. Several countries borrowed heavily in the 1960s and 1970s, ostensibly to jump-start their economies, but instead were trapped in over-indebtedness and underdevelopment. Unable to service their debt, they suffered a debt crisis, a theme to which we shall turn later in this chapter.

13.1.3 Accumulation and Growth

If under-savings handicapped poor nations from investing for the future, over-savings proved to be an equally harming pattern by preventing adequate consumption for the present. Initially, rapid capital accumulation in the planned economies seemed to be successful as vast resources were devoted to the industrialization effort especially in the Soviet Union. However, savings levels and capital accumulation became so excessive that consumption was severely curtailed. Soon commodity shortages of basic goods became so common that provoked insurmountable social problems and making the acceleration of production to backfire. Economic theories spotted this dual inefficiency of under-savings and over-savings and devised rules that would optimize capital accumulation and consumption.

An alternative strategy focused on the expansion of the domestic economy as the mechanism to accumulate production factors. Under this model, the goal was to create economies of scale that will allow the country to create its own national industries and achieve the required threshold internally. The model became popular in the post-war years in Latin America as will be examined in more detail in the next section.

In other countries, economic development was by and large the result of a "quantitative expansion" of production factors. This is the case of the exponentially high growth recorded in the so-called "Asian tigers" in the 1990s and in other emerging economies. Eventually the rapid capital accumulation came to a halt, either because of diminishing capital returns or the labour force ceasing to expand any further, fully confirming the predictions of the Solow-Swan model.[2]

Other nations, however, were able to record high growth rates over long periods, which could not be explained by ever-rising capital or labor. To explain this, a third component was introduced in the growth equation that was acknowledged as the "Solow residual". This accounting of economic growth became prominent in the 1970s, with the revolution in technology and computers. The residual factor could sustain high growth rates even after traditional forms of accumulation had run out

[1] The Marshall Plan provided approximately 13 billion dollars for European reconstruction. Including other aid programs, total financial assistance provided by the United States after the war represented about 10 % of American GDP. For a description see Wexler, *The Marshall Plan Revisited: The European Recovery Program in Economic Perspective* (1983).

[2] For a revealing description of this problem see Krugman, *The Myth of Asia's Miracle* (1994).

of steam; an economy could overcome the problem of diminishing capital returns, since new technologies allowed increasing total factor productivity.

Paul Romer (1986) launched a new generation of economic theories based on the concept of "endogenous growth", according to which the accumulation of production factors is influenced by a series of institutional and behavioral characteristics.[3] Previous experience in factor specialization and accumulation, investments in technology and human capital, a solid institutional framework, can all multiply the effectiveness of the factors of production and ensure sustainable growth rates.

13.1.4 Income Inequalities and Distribution

Apart from economic growth and the ranking of nations in the development league on the basis of per capita income, the issue concerning domestic income distribution started attracting attention in economic theories. Hitherto, Pareto's law of 80/20 distribution between the rich and the poor seemed a natural confine of what policy making could attain. Higher GDP growth was not by itself sufficient to improve allocation, and would more likely attenuate disparities. In fact, in the 1950s Kuznets (1955) had already identified a tradeoff between rising per capita GDP and inequality levels: when an economy grows, disparities become initially more profound, since citizens have unequal access to new opportunities. Only after participation rises and a certain average income is attained, inequality gradually decreases. This hump-shaped relation became known as the Kuznets curve.

Swedish economist Gunnar Myrdal refused to accept both the Paretian claim on the constancy of the gap and Kuznets' conclusion that economic development leads to an initial increase of inequality. Instead, he argued, societies should instead make sure, through education and the provisions of the welfare state, that citizens are dully equipped with the necessary qualifications to board on the train of economic development from early on. If successful, inequality will decrease simultaneously as average prosperity increases.

13.2 Capital Controversies

In the 1950s, two Keynesian off-springs contested with diverging interpretations regarding the role and appropriate measurement of capital. Proponents of one theory rallied around Nicholas Kaldor, Joan Robinson and Piero Sraffa, all with the University of Cambridge, England. Adopting Wicksell's propositions, they argued that the value of capital incorporates previous returns; measuring this value cannot, therefore, be solely based on present returns, such as the profit margin, but on the technology that is used in each case.

[3] For an analysis on increasing returns see Barro and Sala-i-Martin, *Economic Growth* (1995).

Choosing between various techniques of production depends on the ability to substitute capital and labour and this process is determined by the *real* rates of return. This led to the concept of "capital switching and re-switching" across different technology profiles.[4] In Sraffa's framework, income distribution does not just depend on technology, as the neoclassical model presumes, but on external factors which include the monetary policy as well as—echoing the Marxist theory—the power balance between capitalists and workers. Technology affects both production and prices, in order that consumption and savings patterns of households can be finally determined.

At the MIT in Cambridge, Massachusetts, Paul Samuelson (NLE 1970) and Robert Solow developed a neoclassical theory under which capital accumulation is determined by current *market* returns. This theory implies that savings are sufficient to kick-start a phase of capital accumulation and economic growth. In contrast, the theory of Cambridge, England, asserts that savings do not automatically determine investment and some mechanism is needed to facilitate this transformation and allow capital accumulation.

The Cambridge capital controversy, as the debate came to be known, was long and heated, but ended in doubt as to how conclusive it was. Though both theories offer useful insights, they also suffer from simplifications. The MIT overarching assumption that capital has the same marginal value throughout the economy is highly problematic, since it is tantamount to assume that equipment can immediately adjust to any technical change.

The Cambridge England approach failed to influence the dominant economic paradigm of the late twentieth century, but post-2008 it might have a second life. Capital returns in Western economies plunged after the financial crisis and a massive disinvestment took place, especially so in Europe.[5] The process of restructuring the capital stock includes the decommissioning and transformation of capital, massive lay-offs, new training programs, even a relocation of workers and businesses to other sectors and places. To accelerate the return of growth, governments can boost public spending in infrastructure projects and facilitate the process of capital transformation.

The US Cambridge School, on the other hand, would be hard-pressed to account for the prolonged duration of the crisis, given that the factors of production, according to the neoclassical model, adjust quickly to change. Possibly a more constructive compromise of the two schools may suggest that, in normal times, capital accumulation follows the path described by the neoclassical paradigm of the US Cambridge theory. In a crisis, however, capital transformation takes time and

[4] For a comprehensive account of capital switching theory, see Burmeister, *Capital Theory and Dynamics* (1980, Chap. 4). See also Screpanti and Zamagni, *An Outline of the History of Economic Thought* (2005, Chap. 11.2.2).

[5] A study published by the European Investment Bank (EIB) shows that the EU is hit by "a historically unprecedented collapse in fixed capital formation"; see Kolev et al., *Investment and Investment Finance in Europe* (2013).

may lead to prolonged imbalances, recession and unemployment, as predicted by the Cambridge UK approach.

13.3 Inward Theories and Outward Crises in Latin America

The issue of economic development in Latin America attracted a lot of attention and influenced many economic theories throughout the twentieth century. Before World War II, most Latin American countries had become big exporters to the US and Europe, mainly in raw materials and non-manufactured agricultural products. An export-led economic boom was set off but dependence on external demand meant that the 1929 economic crash and subsequent world recession hit their economies. Nevertheless, as they did not become embroiled in the war and maintained amicable relations with all fighting countries, they were able to sustain some economic growth and build substantial foreign exchange reserves.[6] After the end of the war, they used their reserves to nationalise privately owned enterprises, while searching for a new model of economic growth, this time based on their domestic market and their own powers.

13.3.1 The Inward Development Theory

To provide theoretical underpinnings for such an endeavour, the School of "Import Substitution Industrialisation" (ISI) emerged and embraced the following prescriptions:

(a) Imposing import tariffs and other protectionist measures. Some imports of goods that were produced domestically were banned altogether.
(b) Providing state subsidies to all industries and sectors producing goods that can compete with imported ones.
(c) Imposing export ceilings on raw material. The goal was to keep natural resources within the country, so that they could be used at a later stage when the economy was to become industrialised.
(d) Promoting regional integration and trade cooperation, in order to create economies of scale that could further facilitate the growth of national industries in the area.
(e) Attracting selective foreign investment only in non-strategic sectors, since the whole idea was to expand these sectors domestically for the benefit of the national interests.

[6] For a comprehensive overview of Latin American economic history in the nineteenth and twentieth centuries, see Bulmer-Thomas, *The Economic History of Latin America since Independence* (2003).

The Economic Commission of Latin America (CEPAL), set up in the UN and headed by Argentinean economist Raul Prebisch, was pivotal in developing and disseminating the ISI model in the region in the 1950s and 1960s.[7] Prebisch had been the president of the Argentinean central bank during the 1929 crisis and was deeply impressed by the fact that international recession had caused the prices of primary goods exported by his country to plummet, whereas the price drop of industrial products imported from the United States was far less dramatic.[8]

This allowed the developed industrial countries to alleviate a more painful adjustment, whereas nations on the periphery suffered a substantial contraction of national income and trade balances deteriorated as international demand weakened. According to Prebisch, the unequal consequences of the crisis could be accounted for by the long term dependence of Latin American economies on overseas and more developed markets. Latin America paid the price of not having national 'value-adding' champions in industry.

Around the same time, British economist Hans Singer, after thoroughly examining international data, had reached similar conclusions on the unequal distribution of benefits from trade; poor nations are the losers, since their exports have minimal added value.[9] Singer proposed a model of "balanced growth" between developed and developing economies. The combination of those views led to the development of what became known as the "Singer–Prebisch thesis" which gave birth to the brand of '*structuralist economics*'. This approach dismissed the dominant neoclassical model according to which economic development is determined by the comparative advantages of each country and by the capital accumulated through the marginal returns mechanism of the neoclassical model.

The United Nations Conference on Trade and Development (UNCTAD) was established in 1964 as the UN's organ responsible for fostering investment and trade integration in Latin America and other developing countries. Prebisch was the first secretary general of UNCTAD, aspiring to disseminate his favourite model of economic development. Governments of almost all political shades (from Argentinean Peronists to Chilean Marxists) adopted the ISI model in an effort to achieve economic development and lessen their dependence on the US.

Structuralist economic theories were gradually enriched with insights from other sciences, including sociology and history, identifying a number of reasons which accounted for the fact that some countries trailed behind others for too long. The two main causes were the history of dependency of the periphery from the Western core, and a blind adoption of Western institutions and practices instead of developing their own models and structures.

---

[7] Bulmer-Thomas, op. cit., Chap. 9.

[8] This happened because industrial products have more value added and lower demand elasticity, thanks to advanced specialization in comparison to the predominantly raw exports from Latin America.

[9] Singer passed away in 2006, at the age of 95. He was such an enthusiastic researcher that at the age of 86 he signed up for a speed-reading course (*The Economist*, Obituary, March 9, 2006).

Although the structuralist approach echoed the Keynesian tenets on insufficient domestic demand and by implication the need for state intervention, it mainly attracted neo-Marxist economists, like Paul Baran, Paul Sweezy and Andre Gunder-Frank. Neo-Marxists underlined the shared, common characteristics of underdevelopment in their aim to provide a unified theory of international exploitation and thus build an alliance of the oppressed to overthrow the capitalist status quo. Instead, Prebisch focused on the 'exceptionalities', i.e. the indigenous structural characteristics each country had to overcome in order to escape dependence. Rejecting the call for a unified anti-Western bloc, his prescription was to achieve economic development through national reforms and country-specific adjustments.[10]

13.3.2 The Outward Crisis

Theories and policies aimed at economic self-sufficiency continued to stir Latin America in the 1970s, though in the 1980s their popularity withered with Prebisch himself denouncing the effectiveness of ISI policies. Latin America economies were facing mounting problems and the main reasons were the following:

a. State subsidies of national industries led to excessive fiscal burdens and bloating public debts.
b. Domestic industries only partly succeeded in substituting imports and many crucial goods (machinery, equipment, know-how, etc.) were still imported. Meanwhile, exports also suffered both because of the raised costs after the oil crises in the 1970s and the fact that trading partners responded by symmetrical protectionist policies. Decreasing export revenues led to swelling current account deficits and balance of payments crises.
c. Protectionism caused dramatic price rises of domestically produced goods as compared to international prices for the same products. Higher wages and rampant inflation followed, hurting competitiveness even more.

At the end, Latin American countries were trapped into twin deficits and high levels of internal and external debt. Their response was to use the only weapon left in their arsenal: increased money supply in order to finance the budget deficit through inflation tax. The result was a stagflation crisis at a multiple. Given the limited capabilities of domestic industries, monetary expansion further fuelled inflation and led to currency devaluations. A vicious cycle was enacted, with the devalued currency increasing inflationary pressures which in turn provoked violent exchange rate movements.

[10] A collection of articles written by Prebisch and Neo-Marxists can be found in Kanth, *Paradigms in Economic Development: Classic Perspectives, Critiques, and Reflections* (1994, Part II).

13.3.3 Currency Crises and Theories of Restructuring

The spike of oil prices in 1973 and 1978 was followed by skyrocketing prices of raw materials and basic food items such as coffee and meat. At first Latin American economies seemed to benefit, as they were big exporters of primary products and foreign lenders had no misgivings to open credit lines. However, the rise of the world interest rates was hard pressing the producing countries to service their debt, and led many of them to default crises, economic collapses and outright replacement of currencies. In several Latin America countries, crises caused not just the overthrow of the economic model but also great social upheavals, political regime changes and hecatombs of victims.[11] In most cases, the ISI model was replaced by exactly opposite policies prescribed by the monetarist orthodoxy and including among others: [12]

a. Tight monetary policy to combat inflation.
b. Fiscal austerity to cut deficits
c. Extensive market deregulation.
d. Drastic wage cuts to improve competitiveness.
e. Quick privatization of strategic sectors.

The new policy paradigm was divorcing stabilization policies from growth initiatives to such an extent that prompted Williamson (1990) to coin it as the 'Washington Consensus' because:

> [a] striking fact about the list of policies on which Washington does have a collective view is that they all stem from classical mainstream economic theory, at least if one is allowed to count Keynes as a classic by now. None of the ideas spawned by the development literature—such as the big push, balanced or unbalanced growth, surplus labor, or even the two-gap model—plays any essential role in motivating the Washington consensus.

The implementation of such policies meant the termination of inward development theories and policy experimentation on self-sufficiency, but not quite the end of crises. As there was no prioritization on how policies should be implemented, their impact was front-loaded causing further recession and a new wave of income inequalities. Joseph Stiglitz (NLE 2001) launched a polemic against the program by arguing that:

> the net effect of policies set by the Washington Consensus has all too often been to the benefit of the few at the expense of the many, the well-off to the expense of the poor.

[11] In her book *The Shock Doctrine* (2006), Naomi Klein argues that the juntas were instruments of neoliberal policies designed by the "Chicago School", under the guidance of Milton Friedman. Friedman categorically denied any ties to the Pinochet regime in Chile, despite praising its economic policy.

[12] In most countries the new model was implemented by brutal dictatorships. An exception was Peru, where a dictatorial regime ran the course of inward development with monetarist policies being adopted after the return of parliamentary democracy.

> In many cases, commercial interests and values have superseded concern for the environment, democracy, human rights and social justice.[13]

In reality, most countries continued to run severe economic imbalances and were soon entrapped in explosive debt processes and a new phase of serial collapsing.

13.4 The Debt Trap of Developing Countries

One of the failures of the post war international system was its lack of success in setting poor countries on a path of sustainable development, economic stability and social justice. At first, credit from advanced economies—especially the US—and institutions like the World Bank appeared to be a policy strong enough to kick-start development of the weakest parts. Ultimately, the plan frequently went wrong for reasons such as the following[14]:

First of all, many poor countries had long been ruled by undemocratic and corrupt regimes. Easy access to credit—agreements signed with much pomp and fanfare by the leaders of Western nations and local tyrants—provided ruthless and kleptocratic regimes with funds they could turn to their own advantage; put in personal offshore bank accounts or buy weapons that were eventually used against rival countries or their own citizens.

Another reason was that development aid often was just a pretext that allowed lenders to establish their geopolitical influence and make poor countries dependent on global superpowers; as for example, the US, Soviet Union and, more recently, China. Favourable lending terms—as compared to market rates—could mean very little for the growth potential and prosperity of recipient countries, if credit was not attached to conditions for its proper use. Without much growth, recipient countries soon found themselves having unsustainable debt levels.

A debt trap set-in gradually: the assumption of aid programs was that cheap credit would be spent by poor nations on infrastructure and other development projects: if all went according to plan, the economic benefit would exceed the cost of repaying the loans, leaving a surplus to the recipient country. This idealistic—or even naive—picture painted by development theories was vastly disproved in reality.

In the absence of growth, interest payments depress an already low national income, triggering economic contraction. If fiscal tightening is attached as a condition to the loan agreement, recession might deepen and disposable income plummets. The cost of servicing the debt increases further, interest rates rise to capture the increasing country risk and eventually the recipient country loses access to the market. The economy suffers from a liquidity crisis, interest rates rise even more, and the debt trap ultimately destroys growth potential.

---

[13] Stiglitz, *Globalization and its discontents* (2002, p. 20).

[14] For a detailed account, see Herz, *IOU: The Debt Threat and Why We Must Defuse It* (2005).

Then the recipient country starts looking for a way out. The most obvious choice is to default on its debt obligations, a decision often accompanied by political regime change. Sovereign default episodes were so frequent in the 1990s that the international community sought to find a solution. The debt relief initiatives of 1985 and the Brady plan presented in 1989[15] failed to solve the problem, which took unprecedented dimensions in the 1990s. More drastic intervention was needed, with generous repayment or even debt absolution programs designed to help the poorest countries.

A moral hazard problem lurked in debt relief efforts, regarding the risk that a debt reduction may simply encourage debtors to borrow more and arrive in the same situation in the future. The risk was real, since many authoritarian regimes that won debt forgiveness rushed to exploit it by getting new loans and spending the money to serve their own interests, funding populist projects and new armament programs.

Economists were challenged to devise a solution to the moral hazard problem and soon a host of conditionality theories sprang off. To avoid the programs being directed and abused by tyrannical and corrupt regimes, a major prerequisite of debt forgiveness was proposed to be the promotion of democracy, fairness, accountability and transparency. It is well established that foreign aid boosts growth, but only in countries with good governance and transparent economic management. Critics, however, argue that political conditionality may backfire if perceived as foreign interference and suggest that aid should be evaluated on a project basis by the criterion of improving quality of life.[16]

Additional precaution may be enforced by reallocation clauses that oblige recipient countries to cut armaments and match the reduction with an equal increase in spending on health and education; this would allow poor countries to focus on building a better future for their citizens instead of planning their mutual annihilation. All these elements could be part of a credible system of international rules that would break with the failed model followed in the past. Besides, the design of institutions so as to prevent future over-borrowing and wasteful spending led to a rich blending of economic theories with political sciences and opened up new research initiatives worldwide.

---

[15] Nicholas Brady was then US Secretary of Treasury. He proposed to convert bank loans to Latin American countries into new, lower valued but tradable bonds. The new bonds were called Brady Bonds.

[16] See Hermes and Lensink, *Changing the conditions for development aid: A new paradigm?* (2001) and for a review of alternative theories Kosack, *Effective aid: How democracy allows development aid to improve the quality of life* (2003). A collection of articles is available in Stokke (1995).

References

Barro RJ, Sala-i-Martin X (1995) Economic growth. McGraw-Hill, New York

Bulmer-Thomas V (2003) The economic history of Latin America since independence. Cambridge University Press, Cambridge

Burmeister E (1980) Capital theory and dynamics. Cambridge University Press, Cambridge

Hermes N, Lensink R (2001) Changing the conditions for development aid: a new paradigm? J Dev Stud 37(6):1–16

Herz N (2005) I.O.U. The debt threat and why we must defuse it. Harper Perennial, London

Kanth RK (ed) (1994) Paradigms in economic development: classic perspectives, critiques, and reflections. M.E. Sharpe, Armonk, NY

Klein N (2006) The shock doctrine: the rise of disaster capitalism. Picador, New York

Kolev A, Tanayama T, Wagenvoort R (2013) Investment and investment finance in Europe. European Investment Bank, Economics Department, Luxembourg

Kosack S (2003) Effective aid: how democracy allows development aid to improve the quality of life. World Dev 31(1):1–22

Krugman P (1994) The myth of Asia's miracle. Foreign Aff 73:62–78

Kuznets S (1955) Economic growth and income inequality. Am Econ Rev 45:1–28

Romer P (1986) Increasing returns and long-run growth. J Polit Econ 94(5):1002–1037

Screpanti E, Zamagni S (2005) An outline of the history of economic thought. Oxford University Press, Oxford

Stiglitz J (2002) Globalization and its discontents. Penguin Allen Lane, London

Stokke O (ed) (1995) Aid and political conditionality. Frank Cass, London

Wexler I (1983) The Marshall plan revisited: the European recovery program in economic perspective. Greenwood Press, Westport, CT

Williamson J (1990) What Washington means by policy reform. In: Williamson J (ed) Latin American readjustment: how much has happened. Institute for International Economics, Washington

Post 2008: Challenging the Foundations of Orthodoxy

14

Abstract

Dominant economic theories were too confident on the efficiency of markets to guide and discipline the private sector, thus they concentrated on providing Governments with advice and rules on how to impose fiscal and monetary prudence. In most developed economies, the lessons paid off: inflation was checked, public debt was put under control and investment flourished. The collapse of communism led to further integration of international markets and a new era of global growth and prosperity seemed to be dawning. Some speculative bubbles appeared and disappeared, but were capable neither of destroying the big moderation of fluctuations nor undermining the prospects of further growth. In such exuberance, explosive external imbalances between over-saving and over-consuming nations went unnoticed for a long time. The global crisis of 2008 was soon reverberated as a crisis in economic theories and several questions on how our understanding of inherent instability can be improved are still pending.

14.1 Rationality and Intertemporal Constraints Revisited

A financial crisis in late 1990s gripped Asia and Russia, leading to bankruptcies and recession, but it was short-lived. In 1998, Europe established the Economic and Monetary Union and introduced the single currency, while in the late 2000s the world economy was shaken by a global stock-market crisis that became known as the dot.com bubble. Since then globalization picked up pace, with the so-called BRIC countries (Brazil, Russia, India, China) emerging as regional economic powerhouses. The United States witnessed another financial crisis in 2008 which crossed the Atlantic and, 2 years later, led to a debt crisis in the Euro Area. Given all these remarkable developments, it came as no surprise that economics faced, all over again, major theoretical challenges and economists vouched to re-examine several of its previously glorified premises.

N. Christodoulakis, *How Crises Shaped Economic Ideas and Policies*,
DOI 10.1007/978-3-319-16871-5_14

14.1.1 The Omnipresent Constraint

Of the many applications of rationality theory, the efficient-market hypothesis and the intertemporal utility maximization of households are the most heavily criticized. In contrast to the Keynesian function, where only current income determines consumption, the permanent income hypothesis of consumption assumes a typical consumer who maximizes lifetime utility according to both current and expected future income under a new perception of constraints. The Keynesian constraint that a household does not spend above its income was expanded to the so-called "intertemporal budget constraint" (in short, IBC), according to which it suffices that just the present value of a household's intertemporal consumption does not exceed the discounted income expected to occur over the entire period.

This expansion did more than providing economics with the trappings of mathematical elegance; it gave answers to many unresolved issues that the static consumption function of Keynes failed to address such as, for example, the tendency of consumption to remain stable even as income changed over time, or the sudden demand spikes caused by some new expectation of an income increase in the future. Another consequence was that it provided the theoretical underpinnings for an explosive growth in consumers' credit, ostensibly kept within the limits of the IBC.

A incident from the 'Roaring Twenties' was picked up to symbolically express the ostracizing of reckless behavior: Charles Ponzi, an Italian immigrant who lived in Boston was sent to prison in the 1920s for his money making scum to borrow to live in luxury, and then borrowing even more to repay previous lenders. The IBC amounted to the dictum that economic agents "do not play Ponzi Games"—nor do their descendants. The same assumption was made with regard to governments: "not playing Ponzi Games" meant that they would always be in a position to service the debt accumulated by previously expansionary fiscal policies.

In theory, if the IBC is respected by both households and the government, then it automatically holds for the national economy as well. In the long-term, countries will end up with balanced accounts, avoiding both the large external deficits that cause currency crises and the soaring public debts. Intertemporal consumption theories became very popular in the 1980s,[1] forming the pillars of modern economic theory; economists subsequently focused on analyzing the conditions that help the upholding of intertemporal income constraints.

[1] Their dissemination was facilitated by the publication of Oliver Blanchard's and Stanley Fisher's book *Lectures in Macroeconomics*, in 1989, which became the standard textbook for postgraduate economics programs.

14.1.2 Constrained Governments

In practice though, the private and public sectors were treated in a highly asymmetrical manner. In late 1980s, many economies came out of the stagflation crisis with bloated public debts created by their pointless expansionary fiscal policies. At the same time governments struggled to revive investment after they saw that high interest rates due to the increased borrowing had simply crowded–out private spending. In these conditions, fiscal tightening policies became the norm, with governments targeting lower deficits and striving to reduce debt levels. The recipe was thought the only way of marrying price stability with sustainable economic growth. To this effect, extensive empirical research demonstrated at that time that cutting government spending was not only compatible with economic growth but even a catalyst for it.

The ISLM paradigm which had shaped the postwar generations of economists into believing in the efficacy of government spending was replaced by a new orthodoxy preaching the benefits of fiscal tightening and monetary stability. The new theories asserted that even short-term fiscal expansion is ineffective, since—as suggested by the intertemporal model—any additional spending is immediately countered by the expectation of additional future burdens, as predicted by the resurrected "Ricardian equivalence" proposition.

Many fiscal monitoring tools were designed to ensure that governments upheld the IBC rule. Though not always successful or effective, they improved over time. Examples include the government borrowing limits introduced in Britain during the 1980s,[2] the balanced budget provisions added to several state constitutions in US and Canada, the Stability and Growth Pact in the Euro Area, the debt brake rule adopted in Spain in 2011, among many others. Furthermore, governments are under the relentless monitoring of international credit rating agencies which assess their ability to service their debt; in other words to uphold their IBC.

14.1.3 Whatever Happened to Household Budget Constraints?

In the private sector, however, no such mechanisms or scrutiny at a comparable detail existed. For example, the 1991 agreement by European governments to form an economic and monetary union set out specific criteria[3] which EU member states are required to meet before adopting the single currency. Almost all criteria were

---

[2] The Public Sector Borrowing Requirement (PSBR) set the ceiling for the maximum budget deficit of each fiscal year. Later, PSBR targets were consolidated into a medium term budget policy.

[3] These are known as the Maastricht criteria. They required governments to achieve low budget deficit and debt levels (below 3 % and 60 % of GDP respectively), low interest rates and inflation, as well as stable exchange rates. Only inflation is indirectly related to the private sector's behavior regarding price setting and wages.

exclusively referring to the public sector, with not a slim attention paid to the behavior of the private sector.

As a result, it was left entirely to the banks to assess the creditworthiness of their clients, by following a newly body of regulations coded as the Basel Accords. Each successive revision of the regulations was made in order to guard against the repetition of a previous crisis, only to prove ineffective when the next crisis hit. Such regulations were deemed necessary after the 1987 financial crisis leading to Basel I, which set out minimum capital requirements for banks. After the 2000 stock market crises, the code of Basel I was amended to the Basel II framework which included tougher rules for bank capitalization to guard against credit risks, but was in turn superseded by the 2008 crisis. Basel III is an even stricter regulatory standard with provisions for bank stress tests and calls for the separation of commercial and investment banking activities.

These repeated failures can be explained in a Principal-Agent framework: the rules set out by the Principal (in this case the Government) were vague enough to thwart the Agents (banks) from flouting them to maximize their own profits. A case in point concerns bankers' bonuses which were contingent on the loans authorized by bank executives, leading to relentless credit expansion and risky leveraging. Several banks obviously ignored the strictures of the IBC theory, and a period of uncontrollable credit growth was unleashed. On the other hand, the globalization process ensured that even more funds were in search of profitable investments, a trend that coincided with a reckless and dramatic rise of private consumption inconsistent with actual progress in productivity or income growth rates. Thus it was only a matter of time when private IBC collapsed worldwide, causing the 2008 crisis.

An alternative theory regarding the 2008 financial crisis asserts that the culprit was again the public sector.[4] According to this interpretation, the financial system was unable to uphold the IBC because Governments were pressing credit institutions to approve mortgages so they gain popularity and votes. According to this theory, regulatory intervention is the root cause of the crisis, while relaxing market regulation and oversight is hyped as the solution. To illustrate this point, proponents of this theory denounce the "too big to fail" slogan, which perpetuated the misallocation of financial capital.

The problem with the above approach is that it seems to confuse pre-crisis and post-crisis periods. It is perhaps useful in describing what needs to be done to avoid reckless behavior before a crisis erupts; if it is enforced after the crisis all it does is to aggravate the problem for households and accentuate economic contraction. This is what happened in the 1930s and is repeated in certain Euro Area economies nowadays. This approach is also blind to another failure of free markets. Even if

[4] Some public commentators rushed to exonerate banks and turn the blame on governments by arguing that "this crisis was not caused on Wall Street—it was caused in the White House"; *The Spectator* (2008, article by D. Sewell). Similar views were expressed by Austrian School economists; see, for example, Balcerowicz, Polish central banker and politician (2010).

governments had not actively supported private credit expansion, bankers would be more than happy to oblige, in order to secure their bonuses.

The principal-agent scheme failed indeed in regulating and monitoring the banks, but this failure was not the result of government policies to support poor households, or at least it was not the prime reason. The scheme collapsed when banks (the agents) failed to serve their depositors (the principals, who ask the banks to safeguard their savings). Banks disregarded their obligations to depositors by using their savings to create high risk financial products. This conflict between the conservative behavior of depositors and the aggressive investment choices of bank executives played a crucial role in the crisis. This is why the latest developments in economic theory have abandoned the "rational agent" paradigm that had overtaken free-market and deregulation enthusiasts[5] and are rediscovering the usefulness of institutional "Chinese Walls" that separate banking and investment activities.[6]

14.2 The Delusion of Perfect Calculation in Capitalism

The rational expectations hypothesis states that the future value of an economic variable can be efficiently predicted, since agents use all relevant and available information on the current and future state of the economy to make optimal forecasts. In the model, the actual future value of an economic variable differs from the one predicted by economic agents only by a term with zero mean. If all errors are considered to follow a similar random process (called "white noise"), then agent expectations are model-consistent and systematically unbiased.[7]

The rational expectations hypothesis assumes that all economic agents have the same access to information and are equally capable of processing it. However, this assumption is anything but realistic and begs the question of why this theory has become so popular. The answer is simple: without this hypothesis all predictions about policy outcomes would be clueless, with an infinite number of possibilities depending on the model used each time to form expectations. It also raises ethical questions about economic policy: if a policy proves to be successful in practice, it would be so because affected agents were unable to anticipate its consequences!

It is therefore very important to have the ability to evaluate an economic policy—even against unrealistic benchmarks—based on rational expectations, and to know how the economy would perform if all economic agents share the same information. To face the implausibility of rational expectations, some more realistic alternatives have been proposed:

---

[5] Alan Greenspan had used the memorable phrase "*irrational exuberance*" during the dot-com bubble, when he was the chairman of the Federal Reserve.

[6] Until the great walls are built, one proposal to mitigate market speculation is to impose a financial transaction tax, as economist James Tobin (NLE 1981) had suggested during the 1970s. The EU Commission recently presented details about the proposed tax.

[7] When the expected value of error is not zero, the expectations mechanism is considered problematic and the expectations biased.

14.2.1 Theories of Imperfect Information

Regardless of the nature and technical complexity of expectations, perhaps no economic model can make error-proof predictions about the future performance of an economy. Even if all economic agents behaved exactly as the model assumed, there is always the chance of a change somewhere, an unforeseen event or a variable changing course for some hitherto unsuspected reason. In 1885, in a prophetic and sarcastic passage, Russian novelist Fyodor Dostoyevsky had warned against those professing to have the power to make accurate predictions about the future of the economy and society:

> . . . Then. . ., new economic relations will be established, all ready-made and worked out with mathematical exactitude, so that every possible question will vanish in the twinkling of an eye, simply because every possible answer to it will be provided. . . Of course there is no guaranteeing (. . .) that it will not be, for instance, frightfully dull then (. . .) but on the other hand everything will be extraordinary rational. Of course boredom may lead you to anything. . . I, for instance, would not be in the least surprised if, all of a sudden, apropos of nothing, in the midst of general prosperity, a gentleman. . . were to arise and. . . say to us all: 'I say, gentlemen, hadn't we better kick over the whole show and scatter rationalism to the winds, simply to send these logarithms to the devil, and to enable us to live once more at our own sweet foolish will?' . . .What is annoying is that he would be sure to find followers—such is the nature of man.[8]

Several approaches were developed to address the unrealistic premises of rational expectations, ranging from one extreme where behavior changes constantly as the agent processes information to the other end where agents are passively accepting developments. A popular alternative is the so-called adaptive expectations model, which states that current expectations are based on some weighted average of past values. In this case there are of course many possible solutions, depending on how much weight is given to the rational expectations component.[9]

Herbert Simon (NLE 1978) presented a model according to which economic agents do not try to find the best option available to achieve optimal outcomes, but are content with 'satisfying outcomes' that will make them happy. Simon (1979) argued that economic phenomena are inherently uncertain and the cost of finding the optimal solution may be far greater than the benefit. If expectations are not satisfied despite having paid dearly for processing them, then the agent ends up being worse off. Cheaper prediction tools are needed, accompanied by measures to mitigate the impact of uncertainty. Simon advises three ways:

(a) Smart heuristics that improve the quality of information.
(b) Measures to make the system less sensitive to shocks.
(c) Keeping alternatives open if choices turn out to be of high risk.

---

[8] Dostoyevsky, *Notes from the Underground* (1864, Chap. 2, p. 17).

[9] According to Joan Robinson, when decisions are based on expectations, the path to equilibrium and recovery after a disruption are two entirely different phenomena. See her book, *Contributions to Modern Economics* (1978, Chap. 12, "History versus Equilibrium", p. 127).

Tversky and Kanhemann (NLE 2002) examined individual behavior and strategy by applying psychological and heuristic methods instead of the complex functions in large-scale economic models. Haavelmo (NLE 1989) has developed probability theory models which allow the quantification of "behavioral economics".

14.2.2 Chaotic Systems

Another reason why the rational expectations hypothesis is often disproved by reality relates to a system's sensitivity to initial conditions. After the Mexico crisis in 1982 and the upheaval in foreign exchange markets, top scientists gathered in the American city of Santa Fe to see if there are alternative mathematical models better able to account for crises than the hitherto used linear systems.[10] They focused on a branch of mathematics known as 'chaos theory', which analyzes non-linear, dynamic systems so sensitive to initial conditions that even infinitesimal deviations can lead to completely different outcomes. It is noteworthy that though a chaotic system may be described by a simple non-linearity, its sensitivity to initial conditions makes the statistical distribution of variables to be unknown; thus, it is impossible to make predictions about future fluctuations and properly evaluate risk.

14.2.3 Experimental Economics

The relatively new branch of experimental economics tries to replicate real-world effects by collecting data on the possible behavior and interaction of individuals and groups. It turns out that participants display a more cooperative behavior compared to the competitive outcomes predicted by zero-sum games. In the case of currency wars, for example, players who represent 'countries' may eventually learn that they have more to gain from a regime of stable exchange rates; in arms race experiments, competitors may conclude that mutual disarmament is preferable. Behaviors appear to reflect a "habitual cooperation", escaping the curse of endless competition and conflict. The tendency of cooperation can be protected from opportunistic actions via sanction mechanisms or a powerful commitment to compliance.

All the above approaches deviate from the sophisticated patterns derived by the rational expectations models and this implies that decisions based on the latter risk to have a hard landing on the reality of actual markets. The 2008 credit crunch is viewed as such an eventuality big scale, and previously acclaimed theories are now anxiously reconsidered.

[10] This meeting is described in Beinhocker, *The Origin of Wealth: Evolution, Complexity, and the Radical Remaking of Economics* (2006, p. 45).

14.3 Inefficient Theories and Markets

A number of asset pricing models in the 1950s tried to formulate criteria for guiding investment behavior. The most well-known is the Capital Asset Pricing Model (in short CAPM), developed by William Sharpe (NLE 1990). It is based on the so-called "beta" ratio of an asset which compares the volatility of an asset to the volatility of the market in general; a high value indicates that the asset is more volatile than average. Thus, the risk in investing on this asset is high, but potential returns are high as well. On the contrary, blue-chip stocks, i.e. companies with a high market capitalization, have a lower beta, meaning they are low risk and bear low return.

14.3.1 The Efficient Market Hypothesis

CAPM theory had determined investment choices for decades, until it was sidelined by a rival approach. In 1965, Eugene Fama (NLE 2013), at the University of Chicago, presented the Efficient Market Hypothesis (EMH), which assumes that financial markets are always informationally efficient: in other words, the prices of traded assets already reflect all available information, such as transportation and preservation costs, the demand curve, etc. In accordance with the theory of rational expectations, the market is efficient in pricing an asset, after processing information in an objective and verifiable way. The CAPM is therefore useless and Fama boasted that his hypothesis was a "bullet in the heart" of CAPM, and, therefore, "beta is dead".[11]

There are three versions of the EMH which correspond to different degrees of efficiency: under the weak-form EMH, current prices reflect all public information and past performance. The semi-strong form claims that current prices reflect all public information and change instantly to reflect any new information becoming available, while according to the strong-form EMH current prices reflect all possible information. If the market price of an asset systematically deviates from the EMH hypothesis, the asset is a "buy" (if the market price is lower) or a "sell" (if it exceeds the EMH pricing).

14.3.2 The Quants: Power Without Theory

'Technical analysis' is a highly complex armor of mathematical tools used for forecasting the direction of prices in the stock market as well as making projections for other economic phenomena. Myron Scholes and Robert Merton (NLE 1997) were the first to apply mispricing identification analysis for guiding investors' decisions. Their reputation, however, was tarnished after a hedge fund they

[11] Quoted by The New York Times, February 18, 1992.

owned was liquidated in the early 2000s, and they conceded that "[the] whole approach was fundamentally flawed".[12]

During the last two decades, "technical analysis" has been used so extensively in financial market projections that its practitioners acquired the reputation of being a computational movement, known as "Quants".[13] Most of them have a strong background in mathematics and physics, and their deep knowledge of stochastic calculus, partial differential equations and non-linear systems allowed them to solve highly complex functions. Ironically, the more complex their models were, the harder it became to check the validity of their assumptions and eventually financial markets looked like automated betting machines—a type of trading robot.[14]

Online market analysis allows the system to identify falling and rising stock prices and then initiating an arbitrage process through market transactions that occur in the blink of an eye. Stock orders in global markets are given automatically, 24 h a day, 7 days a week. An example is their "closing bell" trading activity: using supercomputers, they analyze the market and give trading orders at the last minute, having the last say on the direction of the market. The methods of Quants looked like occult practices to the average economist: the justification was that only in this way the true nature of economic phenomena is reflected in their prices, which vary depending on the changing conditions. By analyzing these variations, they can come up with a probability distribution which is the closest possible to the actual outcome.

The Quants revolutionized the financial sector. Some envied them for the mythical returns they achieved with their complex derivative choices, but many hold them as the main culprits of the 2008 financial crisis by fueling the reckless expectations of investors. The models used by Quants became totally unrelated to the economic experience or even to mere experiments that used to test an economic theory. Given the lack of a theoretical foundation and the complexity of the mathematical calculations they use, their prescriptions are essentially a "black box". As happens in aircrafts, the black-box of the financial markets was dutifully scrutinized after the crash.

14.3.3 From Mandeville to Mandelbrot

A leading mathematician, Benoit Mandelbrot (1924–2010), challenged the ability of technical analysis to describe and predict market outcomes. According to

---

[12] Quoted by Mandelbrot, *The (mis)Behavior of Markets: A Fractal View of Risk, Ruin and Reward* (2008, p. 107). The firm dissolved after the collapse of the Russian bond markets in August 1998. Federal authorities spent 3.6 billion dollars to prevent a fallout.

[13] From the word "quantitative".

[14] Patterson, *The Quants* (2010, p. 138) describes how a band of Quants developed a trading robot aptly called "Midas". Within 5 years, it had made one billion dollars for the bank hosting it. Competitors lost and accepted the fatal advice "always trust the machine".

Mandelbrot, prices do not always confirm the EMH and the error margin in projections seldom is zero. The normal distribution which characterizes EMH is "leptokurtic", conveniently implying that extreme mispricing episodes are very rare, but if errors follow more "platykurtic" distributions the likelihood increases. Mandelbrot wonders why, despite their obvious weaknesses, these analytical techniques are still in use. The reasons are:

> Habit and convenience. The math is, at the bottom, easy and can be made to look impressive, intractable to all but the rocket scientist. Business schools around the world keep teaching it. They have trained thousands of financial officers, thousands of investment advisors... It is false confidence, of course. The problem lies at the root of the standard model, in its assumption that the best way to think about stock markets is as a grand game of coin flipping.[15]

The most impressive finding by Mandelbrot is that complex phenomena can be the aggregate outcome of elementary events. Based on his theory of fractals, Mandelbrot argued that even simple phenomena can, when piled up, acquire complex and not always identifiable characteristics. To illustrate this, he offered the examples of mountain ranges, coastlines or of a random walk.

These low-probability but nevertheless significant events are often called "black swans", an idiomatic phrase originally derived in Roman times to express the fact that someone is claiming the impossible. But after Dutch seafarers did discover black swans in Australia in the seventeenth century, the term has been used to characterize improbable but not impossible events. Since the black swan is considered an ugly bird, the term is used pejoratively.[16]

In economics, Mandelbrot's theory implies that an individual agent (a household or an investor) may be rational, but when the number of agents multiplies, rationality may be lost rather than being aggregated. In fact, Mandelbrot turned Mandeville's optimistic eighteenth century vision upside down: the latter had hoped that individual irrationalities will be filtered out by the market into a collective rationalism, and now it is shown that individual rationalities may result in being collective follies.

There might be a solution to this conundrum. Just as the world of Mandeville needed a free market to balance conflicting individual goals, the world of Mandelbrot needs market regulation to preempt the explosive combination of such goals. Without monitoring, the rationality of the individual may result in reckless market irrationality, confirming the fears crisis master-teller Charles Kindleberger had expressed on contagious events.[17]

---

[15] Mandelbrot (2008), *op. cit.*, p. 105.

[16] Taleb's book (2010) on the economic consequences of improbable events is aptly titled *The Black Swan*.

[17] Kindleberger, *Manias, Panics and Crashes* (1978, p. 32).

14.4 Economic Imbalances: Are States Playing Ponzi-Games?

After the 2008 financial crisis everyone agreed that there would be ramifications for the real economy, but there was no consensus on whether the crisis was caused by the imprudence of banks or by imbalances in the real economy.

14.4.1 The Role of Leverage and the Modigliani–Miller Theorem

One consequence of the ineffective regulatory framework was the excessive "leverage" of banks, i.e. the loans-to-their capital ratio. Banks with high leverage become vulnerable to economic downturns, when borrowers fall behind their obligations. Banks with low leverage are conservative and avoid high risk investments.

Leverage has been a central concept in modern economic theory, especially after the work of Franco Modigliani and Merton Miller (NLE 1985 and 1990 respectively). According to their famous theorem, the value of a firm is unaffected by leverage levels, since stock prices already reflect the market's valuation of the risk involved. The hypothesis became very popular in the 1960s and gave the green light to banks to stop being conservative and expand their portfolios without risking the value of their stock.

The Modigliani–Miller theorem (M-M) was conveniently interpreted this way by a financial system eager to maximize its profits from unbounded credit expansion. No one seemed to care that the M-M theorem is valid only under strict—and mostly unrealistic- conditions as they include perfect competition, market clearing prices and perfect information, all applied full-scale and simultaneously. In this respect, the M-M theorem is similar to the Walrasian general equilibrium model which is very insightful as an analytical tool, but its usefulness lies not in its connection to reality but in identifying the distance from it. In short, the M-M hypothesis is a benchmark, not an applicable principle.

Leverage proved to be a very critical factor, both during an economic boom and in times of recession, but not in the direction of gratifying the M-M hypothesis. One could reasonably assume that leverage should be countercyclical—i.e. contracting when the economy grows—to prevent overheating, and increases in times of recession to boost investment. Instead leverage was in effect procyclical, increasing as the economy expanded and fueling a reckless credit boom that made the eventual collapse all the more painful.[18] In practice, even investment banks which should have known better on how the risk is spreading, adopted an extremely procyclical behavior whereas traditional commercial banks were more conservative.

Leverage becomes even more procyclical in a recession because, in order to deal with bad loans, banks are obliged to decrease their loans-to-capital ratio. The

[18] For an account of the impact of procyclical leverage in the recent crisis see Beccalli et al., *Leverage Pro-Cyclicality and Securitization in US Banking* (2014).

denominator can be increased either by raising capital from the market (though this is hard to do in a crisis, when investors are weary) or by state intervention (this is also difficult given that highly indebted governments may resist taking on even more debt). Another way is to decrease the numerator by limiting credit to companies. This is the process of "deleveraging", which obviously accentuates recession.

14.4.2 Imbalances in the Real Economy

Though very few doubt that the financial sector was the epicenter of the crisis, the fact is that its causes should be traced deep in the real world economy. Dramatic imbalances had been previously witnessed in the global economy, eventually leading to the collapse of the banking system. The most important of these imbalances were the following:

(a) During 1998–2007, the US economy witnessed a period of excessive private and public sector consumption, leading to the problem of the "twin deficits", i.e. the co-existence of high current account and budget deficits.
(b) Emerging markets also ran record-high current account deficits during the same period. Running those deficits was partly justified, as they attempted to attract investment, support domestic production and accelerate economic growth. Still, deficits ultimately became unsustainable, credit ratings downgraded and investors were thwarted from further financing.
(c) The above imbalances are symmetrically related with the excessive savings in many countries, including the United Arab Emirates and the BRIC economies. Despite their surging production and income in the previous decade, domestic consumption failed to take off as several of them still apply spending controls or impose religious or ideological patterns.
(d) A third factor that preceded the outburst of the crisis was rampant speculation in oil markets, which became contagious with commodity and food prices skyrocketing as well. The dramatic rise of oil prices was primarily due to speculative moves, as investors were confident that demand for oil would continue to grow and rushed to take positions.[19]

With the western economies consuming more than they could produce and emerging economies producing more than they were allowed to consume, it was only a matter of time before a crisis would occur.[20] During 2001–2007, liquidity was in abundance in global markets but when the international financial community

[19] For an informative exposition of the 2008 oil and commodity bubble see Khan, *The 2008 Oil Price "Bubble"* (2009).

[20] For an analysis, see Davies, *The Financial Crisis*, Chap. 3: "The Savings Glut: Global Imbalances" (2010, p. 17).

finally realized that imbalances are no longer viable and a painful adjustment is just around the corner, liquidity all but dried up.

14.4.3 Wiser, but Not Yet Sufficient

A dramatic and heated debate erupted in the academic community on how policymakers should have acted. Consensus was hard to reach, given the numerous and often conflicting interests involved in the debate. Competing interpretations and suggestions relating to the 2008 financial crisis abound by supporters of Keynesianism, monetarism or Marxism. Many Keynesians saw it as a revival of the Great Depression and proposed a twofold policy of (a) government intervention to kick start the economy and save the banking system and (b) stricter market regulation to prevent the repetition of reckless past behaviors.[21]

Monetarists insist on the merits of a combination of tight monetary policy, wage cuts to bring businesses back to black, and full market deregulation to allow the market to find its optimal state without the harmful and distorting interventions of the state. According to neoliberal economists

> The reasons for the current recession are pretty straightforward: it is hard to get much done if you are scrambling for cash and the normal way of doing business just fell apart. Now that the short-term credit crunch is over, the recession seems likely to be followed by a quick recovery, at least if the government does not get in the way with too many counterproductive attempts to fix things.[22]

Schumpeterian economists have taken the evolutionary approach envisaging new constructive opportunities created by the destruction.[23] Once more, Marxist economists assert that the 2008 crisis has confirmed the inherent weaknesses of capitalism, though this time their conclusions stop short of predicting its imminent downfall. Marxist-leaning politicians asked for the nationalization of the banking system and big corporations, seeing it as a first step towards a planned economy.[24]

A political response was more swiftly reached: governments would bailout problematic banks, while also taking measures to stimulate economic growth and avoid a repetition of the 1930s tragedy, when the financial crisis met no resistance and spread to the real economy causing a global recession. In the 1930s, it had taken at least 5 years to exit the trap of recession. This time, governments intervened promptly, the crisis was kept under control and by 2010 the US and Europe had returned to growth; proof that something was learnt from the crisis and its tragic

---

[21] For example, Krugman, *The Return of Depression Economics and the Crisis of 2008* (2009).

[22] In the words of Cochrane, *Lessons From the Financial Crisis* (2010).

[23] For example, Hanusch and Wackermann, *Global Financial Crisis: Causes and Lessons: A Neo-Schumpeterian perspective* (2009).

[24] For example, by the leader of the third party in Germany (*Die Linke*); see *The Economist* ("What would Marx say?", 15 October 2008).

aftermath.[25] The relief was short-lived, however, and soon the crisis relapsed into the sovereign debt crisis in the Euro Area.

In the aftermath of the 2008 financial crisis, the Euro Area appeared to be an anchor of stability in the midst of global turmoil. As opposed to their Anglo-Saxon counterparts, European banks had been more conservative, avoiding the speculative, high-risk investments in toxic assets. European debt and deficit levels were also lower compared to UK and the US, since the latter had to approve mammoth rescue programs to bailout their banks and ailing corporations. Several non-Euro economies were wondering how much better things could have been, had they joined the single currency. But despite all these early signals, it was the American economy that managed to exit the slump earlier and grow faster while the Euro Area plunged into a debt crisis and recession.

The reason of the crisis can be traced, once more, to the proportions of scale and factor freedom. The Euro Area was created primarily to put an end to currency wars when trade imbalances occurred. Before the single currency was launched, a country could devalue its currency to become more competitive. Sooner or later this would provoke the same response from its trading partners and competitors, leading to mutually destructive currency wars and rampant inflation. The introduction of the common currency made the economies to function in a much larger scale but at the expense of loosing monetary independence. Alternatively, other types of freedom were introduced to allow mobility of capital and labour.

That was in full compliance with the theory of Optimum Currency Areas, according to which if current account deficits become unsustainable, production factors need move to another country to equilibrate the system.[26] In practice, factor mobility did not work according to expectations. On one hand, investment capital was allocated asymmetrically with productive investments flowing mostly in the northern Euro Area countries and real-estate acquisitions taking place in the south. On the other hand, labour mobility was marginal due to a variety of frictions in European labour markets and cultures. The allocation asymmetry was producing vast intra-European imbalances, eventually leading to the credit crunch and the ensuing debt crisis. The Euro Area currently faces a multitude of challenges on how best a currency union can work without a complete fiscal federation and a new wave of theories may come out to deal with.

The crisis once more was proven to be the litmus test for the dominant paradigm and the promised land for new aspirations. One just hopes that the competition unleashed by the crisis on the suitability and wisdom of alternative approaches may help a new and more sustainable paradigm to be formed and subsequently prevail.

[25] The sense of having learnt from the mistakes of the 1930s was epitomized in 2002 by Fed's then Governor famous remark "we won't do it again". Though finally the 2008 crisis was not avoided, Fed's policy was way more effective in containing the crisis effects; see Bernanke, *On Milton Friedman's Ninetieth Birthday* (2002).

[26] Mundell, A Theory of Optimum Currency Areas (1961).

References

Balcerowicz L (2010) How to prevent another global financial crisis? Mimeo/Eliamep, Athens
Beccalli E, Boitani A, Di Giuliantonio S (2014) Leverage pro-cyclicality and securitization in US banking. J Fin Intermed. doi:10.1016/j.jfi.2014.04.005
Beinhocker E (2006) The origin of wealth: evolution, complexity, and the radical remaking of economics. Harvard Business School Press, Boston
Bernanke B (2002) On Milton Friedman's ninetieth birthday. Speech at the conference to Honor Milton Friedman, University of Chicago, Chicago, IL, November 8
Blanchard O, Fisher S (1989) Lectures on macroeconomics. MIT Press, Cambridge
Cochrane J (2010) Lessons from the financial crisis. Regulation Winter 2009–2010 (Cato Institute), pp 34–37
Davies H (2010) The financial crisis: who is to blame. Polity, London
Dostoyevsky F (1864) Notes from the underground. Penguin, London
Hanusch H, Wakermann F (2009) Global financial crisis: causes and lessons: a Neo-Schumpeterian perspective. Institut fur Volkswirtschaftslehre, Beitrag Nr. 303
Khan M (2009) The 2008 oil price "bubble". Peterson Institute for International Economics, Working Paper, no. PB09-19, August
Kindleberger C (1978) Manias, panics and crashes. Wiley, Hoboken
Krugman P (2009) The return of depression economics and the crisis of 2008. W.W. Norton, New York
Mandelbrot B (2008) The (mis)behaviour of markets: a fractal view of risk, ruin and reward. Profile Books, London
Mundell R (1961) A theory of optimum currency areas. Am Econ Rev 51:509–517
Patterson S (2010) The quants. Random House, London
Robinson J (1978) Contributions to modern economics. Basil Blackwell, Oxford
Simon H (1979) From substantive to procedural rationality. In: Hahn F, Hollis M (eds) Philosophy and economic theory. Oxford University Press, Oxford
Taleb N (2010) The black swan: the impact of the highly improbable. Penguin, London

Is There a Methodological Crisis in Economics? 15

Abstract

The key lesson of the global crisis is that there is no such thing as inherent stability in economic systems and only a continuously revised and self-critical theory can provide knowledge, guidance and warnings on how a crisis can be dealt with or avoided altogether. Yet, the propagation of economic ideas in general and the teaching of economics in particular, are frequently characterized by overconfidence and lack of self-critical awareness. Some approaches are indulging in complex calculations disregarding the realism of the assumptions on which they are premised, while other schools of thought preach their ideas, no matter how they come to terms with reality. But economics are going to enjoy neither the deterministic certainties of natural sciences, nor the aloofness of philosophical purity. It is destined to be a knowledge system intrinsically linked to social dynamics and, therefore, continuously challenged by unforeseen problems. A major inspiration for improving this knowledge is by thoroughly reading the history of ideas and how they shaped, and were themselves triggered by, economic history.

15.1 Problems in Teaching Economics

The way that economics is taught in several universities is often characterized by lack of objectivity and realism. In the event of a contradiction with reality, it may be difficult to recognize limitations and shortcomings, so that it is timely revised or abandoned all together. Two reasons associated with such pitfalls are discussed below.

N. Christodoulakis, *How Crises Shaped Economic Ideas and Policies*,
DOI 10.1007/978-3-319-16871-5_15

15.1.1 Kepler's Lesson Forgotten

A constant problem with economic theories is how well they reflect real world trends and developments, no matter how rigorously they are formulated. A characteristic example is the Ricardian equivalence proposition, which asserts that an expansionary fiscal policy is absolutely neutral, because households immediately increase savings to pay for future taxes and preserve their consumption. When governments run enormous fiscal deficits, tax hikes become inevitable and this obviously leads to lower private consumption and increased savings. Despite its theoretical elegance, the Ricardian equivalence falls apart in the case of poor households with no spare income to save anyway, or rich households for which consumption levels are not affected by additional tax burden.

Pitfalls often make the proponent of a theory to omit inconvenient realities. Surely, economics is not the only science suffering from this problem, and Karl Popper has memorably criticized those who believe they hold the keys to absolute knowledge and harmony: from the naturalist philosophers of the sixteenth century who believed that everything is explained by the laws of mechanics, to the proponents of idealism who tried to discover the "elegant harmony" of events and held doctrines based entirely on ideas.[1]

A formidable example of how an otherwise appealing theory can still be uncoupled from reality was Kepler's (1571–1630) discovery that planets move in elliptical rather than cyclical orbits. Like Copernicus previously, Johannes Kepler was convinced that planetary movements follow the "mathematical harmony of the circle". When he made the opposite discovery, at first was reluctant to announce it. But finally Kepler resisted the temptation of personal gratification and, instead of transfiguring reality to suit his mathematical model, defended the accuracy of observations and published the discoveries in 1619. He did not hesitate to change his own assumptions, stating that "harmony does exist in the universe, and is determined by the motion of celestial bodies, but not in the way I had originally imagined".[2]

15.1.2 Relativity and Selectivity

In economics, as in many other sciences, there is no single, general theory that explains all phenomena. Usually, a theory is good at explaining some cases, less so at explaining others or even totally inadequate. It is important that the strong as well as the weak points of each theory are known before it is adopted into a study.

Keynesian theory, for example, cannot deal with supply-side shocks, as it focuses entirely on the side of demand. Marxism cannot account for the resilience of capitalism because it is founded on the hypothesis that capitalism is in a state of

---

[1] Popper, *All Life is Problem Solving* (1999, Chap. 15, pp. 341–49).

[2] A discussion of these issues can be found in Reichenbach, *From Copernicus to Einstein* (1970).

perennial crisis. Monetarism cannot interpret the liquidity trap that engulfs an economy after a recession, because it asserts that monetary policy is an omnipotent tool.

In such cases, theories may be laden with several presumptions that prevent objective teaching and unbiased assessment of a theory. After all, professors of economics were themselves learning and passing judgment on the value and usefulness of the theories long before they were asked to teach them. Ideological bias can be mitigated as long as teaching proceeds by:

(a) Setting a straightforward and clear framework of teaching objectives.
(b) Present competing interpretations by other schools of thought.
(c) Provide a comparative evaluation of alternative theories.

Students of economic theories would immensely benefit from understanding not only the confirmations but also the failures of each theory.[3] This would develop their critical thinking and motivate them to search for new theories and compare existing ones. It would also help future economic theorists to become moderate, open-minded and more open to criticism.

15.2 Competing Theories

Making the proper choice of a theory has been investigated in depth by the scientific community, sparking heated debates between proponents of competing approaches.[4] Below are some of the criteria that could be used in the selection process:

15.2.1 Simplicity

A commonly used criterion in deciding which theory to follow is to apply Ockham's razor and select the simpler one, when more than one are equally good at explaining the behavior of a system. The rule is picking the one which is easier to understand, simpler to use and more transparent in identifying weak points. According to Olson,[5] a good theory is "parsimonious"—it explains a lot without

[3] The most typical expression of the view that a universal theory can explain everything can be found in Stalin's Economic Problems of Socialism in the Soviet Union: "What is needed, therefore, is a textbook which might serve as a reference book for the revolutionary youth not only at home, but also abroad... it must contain everything fundamental relating both to the economy of our country and to the economy of capitalism and the colonial system".

[4] A concise presentation of the methodological issues in economics can be found in the classic book by Blaug, *The methodology of Economics: or how economists explain* (1980).

[5] Olson, *The Rise and Decline of Nations: Economic Growth, Stagflation, and Social Rigidities* (1984, pp. 12–13).

unnecessary assumptions and obscure formulations. Olson adds that the credibility of a theory rests not only on the sheer number but also on the diversity of phenomena it can explain. A model theory in this respect is Darwin's, since it can explain the most diverse cases of evolution, from the smallest activity of bacteria to the size of whales and human beings.

15.2.2 The Test of Changing Conditions

Another criterion used to compare and evaluate competing theories is how good they are in describing the consequences of some unforeseen change. In natural systems this is not as challenging as it sounds, because they remain unchanged in time and the domination or a rejection of a theory does not transform it. For example, it may have taken centuries to explain the nature of light, the composition of water or the DNA mechanism, but as theories came and went the underlying structures remained the same and eventually scientists were able to find the correct interpretations. The same criterion has limited chances in economics, above all because a theory may itself influence the structure and evolution of the system.

15.2.3 Keeping Score

A theory's predictive power (or lack thereof) cannot be established by just one correct or incorrect prediction—it has to be tested multiple times. For obvious reasons, the more frivolous a theory, the less willing it is to be exposed to this test. To ascertain the predictive qualities of a theory, the one invoking the theory should also present its track record by creating a score index of the number of correct predictions as a percentage of total ones. This criterion reveals that many self-proclaimed prophetic theories have very low scores indeed. But even if the score value is high, this does not necessarily mean that the theory is correct and a more fundamental examination of the theory should be undertaken.

It should be noted that, for psychological reasons, the outcomes of predictions are recorded asymmetrically in our memories. When a disaster does happen, everyone remembers the person who had predicted it, with a mixture of feelings of fear or admiration. If nothing happens, very few bother to remember those 'doomsayers", being content with the fact that their prophesies did not materialize. Even if the same person makes a new prediction in the future, very few will remember his past failures.

15.2.4 Normative and Positive Theories

A positive theory is only interested in descriptive and factual statements about the world without taking into consideration how the world should work. The Real Business Cycle (RBC) theory, for example, investigates economic fluctuations but apparently says nothing about how an economy can avoid them. Normative theories on the other hand are interested in the optimal allocation of economic resources to achieve some predetermined economic goal. In other words, normative theories say how things should work, while positive theories describe why and how things actually work.

Historically, many economic theories have been purely normative—they were developed either to justify how the economy worked or to suggest an alternative compatible with other value-based systems (moral, philosophical, political, etc.). In many cases, of course, theories combine the two elements and explain a phenomenon in a way that justifies the theory's premises and suggestions. For example, Keynesianism favors government intervention, whereas neoclassical theory argues in favor of individual and market freedom.

But even some apparently positive theories may have deep normative implications. For example, the RBC theory by asserting that fluctuations are unavoidable responses to exogenous changes implicitly suggests that the government should do nothing in times of recession. Milton Friedman was acclaiming positive economics as being "independent of any particular ethical position or normative judgments", but nevertheless admitted that their conclusions "seem to be, and are, immediately relevant to... questions of what ought to be done".[6]

15.2.5 Economic Agents and Roles

Economic theories can be examined from another angle—whether they are assuming "individualism" or "holism" as the pattern of economic behavior. "Individualism" focuses on the behavior of individual members, while "holism" is focusing on collective behaviors. In economics and other social sciences, specific individual behaviors can lead to different holistic ones.

A problem of this type emerges with intertemporal consumption theory that entails the "representative consumer" whose preferences and rational choices are believed to represent the behavior of society at large. The representative household thesis ignores the variety of household preferences and constraints; for example, poor households operating under liquidity constraints (as in Kalecki's model in Chap. 9) and run into bankruptcy when banks cut them off. Intertemporal consumption theories should develop additional models dealing with the multitude of preferences, existing liquidity constraints and real-life risk profiles. This expanded system will be more sensitive to disruptions, and therefore better equipped to deal with crisis and collapse.

Another issue relates to the multiple roles that are attributed to economic agents but by following a one-by-one process. Economic theories usually ascribe a single specific role to the agents of under examination: in the theory of consumption individuals are consumers who want to maximize their utility; in the theory of

[6] Friedman, *The methodology of positive economics* (1979, p. 10).

employment, individuals are members of a unified labor force; in the public choice theory, individuals are voters with varying preferences, while in financial analysis they are competitive investors. The individuals, nevertheless, may be the same persons exercising many roles, but a unified theory able to describe their multiple roles is still missing.

15.3 How We Read Economic History?

Throughout the journey into the history of economic theories and events, a central question has arisen: how should one read past episodes and benefit from history in general and economic history in particular? Anyone wishing to use historical analysis to defend a theory should, before anything else, fully appreciate the risks of the venture in the hope of minimizing them. Then, one should make such pitfalls known to the reader, allowing them to grasp why a specific historical reference was chosen. Among the most common risks of historical analysis in economics are the following:

When we study the economic history of a period, a question arises about whether some other economic approach—available at the time or not—could have worked better compared to the one that actually prevailed. To answer this question one naturally presupposes a framework of comparing and judging economic outcomes of the period under investigation. For example, do we evaluate outcomes based on Pareto's criterion if some people become better off and no one becomes worse off, or based on a Benthamite collective utility function including only the aggregate improvement for society?

Such an endeavor presupposes an even more arbitrary assumption: that the criterion we decide to use is indeed appropriate for the period under examination. We assume, in other words, that the people who lived in that era would be willing (if could be asked) to be judged by our modern standards. Our retrospective analysis may be reasonable if our criteria refer to fundamental and universal values about liberty and survival. But if we apply other modern-era criteria—for example faster transportation or job mobility—it is by no means certain that the theoretical exercise has any relevance or meaning when applied to societies of the past.

Beyond these preliminary caveats, there is the perennial debate concerning the appropriate use of history. Echoing the respective debates in history sciences, we will now examine the arguments for and against the reevaluation of events in economic history and the history of economic theory. The most common approaches are the following:

15.3.1 Historical Cyclicality

Economists frequently have predetermined views about how a past specific economic phenomenon should be analyzed. At university, one is taught a lot of economic theory before having a chance to discuss real historical cases and events.

Even then, these events are presented in a brief and one-dimensional way as proof of the theory under discussion. It would be much better if similar case-studies of multiple past eras and geographical regions were grouped together and presented, and then test the theories on these phenomena. To give an example, it would be preferable to discuss how various policies dealt with all recorded cases of famine and then compare the various theories of agricultural production, distribution, etc.

More often than not, economic history courses focus almost exclusively on the university's home country historical record. They also typically focus on more recent events, because data and accounts are easier to access. But even when the focus is on national economic history, only rarely do courses examine all the important events or comprehensively analyze all available economic data. The standard practice is for the teacher to handpick specific historical events that "fit" the preferred economic theory, giving the wrong impression that history repeats itself in a cyclical process. Anything that does not fit the theory's "historical laboratory" is frequently ignored.

15.3.2 Historical Necessity

According to this approach, historical processes have happened by necessity. Initially, causes were attributed to the will of Gods; later, to the Church or to secular powers representing God on earth. Hegelian idealism believed that there are laws of history at work, leading to an inevitable path of historical progress. For Marxists, class struggle is the catalyst and engine of the equally inexorable march of history.

Such an approach makes a dubious assumption that there is some sort of "end" to which history is inextricably led. Individuals may be oblivious to this and could not alter the course of history even if they had wanted. The teleological approach allows self-proclaimed Messiahs or groups to claim to be the agents of this historical process. The dead-end is that this theory eliminates the role of individual freedom. All that is left for individuals to do is accept the deterministic stride of history and play along.[7]

15.3.3 Historicism

According to this approach, the study of history should be based only on the actual events, rejecting historical teleology and deterministic interpretations. In economics, historicism dismisses the comparative analysis of different eras and societies, asserting that each epoch is the product of very specific and unique processes, which render it unrepeatable and incomparable. Karl Popper rejects historicism, arguing that it deprives individuals and societies the freedom to shape their own

[7] A thorough discussion of these issues can be found in Ferguson, *Virtual History* (2011, p. 39).

future and adopt, should they desire, an alternative model of social and economic organization.

In turn, Marcuse criticizes Popper's extremely negative approach[8] and argues there are two separate trends in the wider tradition of historicism. Cyclical theories assert that human nature remains always the same, despite superficial societal changes, which leads to an endless repetition of the past. The other trend accepts the existence of general laws in historical progress, but their manifestation is influenced by the free will of people and the existence of the material conditions for change (or lack thereof). It is therefore pointless to foresee with certainty what will happen in the future, but is useful nonetheless to examine the consequences of each alternative path.

15.3.4 Cliometrics

Economists Robert Fogel and Douglas North (NLE 1992) developed innovative contributions and ideas of studying economic history that eventually led to a new branch of economic research called "new economic history" or Cliometrics. The term comes from the combination of the name Clio, the muse of History, and econometrics. North examines the influence of institutions on the economy and, despite his Marxist roots, turns the major Marxist tenet of historical materialism on its head.

Fogel, who mainly accepts the neoclassical theory of production and consumption, focuses on the role of technology and how it affects the productivity of capital and labor. According to him cliometric research should be conducted in a way that is "detached, unemotional [and] respectful of fellow investigators". The approach soon became popular, but also attracted a lot of criticism. Boldizzoni (2011) rejects the rigid conceptual framework that is either explicitly or otherwise assumed in cliometrics and calls for "[e]conomic history [to] lift itself out of the difficult situation it is now in by becoming involved with the genuinely 'social' sciences and with all those scholars who are interested in an innovative interaction... without imposing any particular point of view".[9]

15.3.5 Economic Anthropology

Similar dilemmas about the best approach to analyze economic phenomena abound also with regard to their geographical dimension. Economic anthropology was developed over the previous decades as an interdisciplinary field, focusing on the economic structures and behaviors in non-western countries, societies and local communities. Methodological questions of the type previously examined are also

[8] Marcuse, *From Luther to Popper* (1983, p. 199).

[9] Boldizzoni, *The Poverty of Clio: Resurrecting Economic History* (2011, p. 170).

present in economic anthropology, but the thorniest issue is comparability among non-western models or between non-western and western ones. From this debate three alternative approaches sprung off, affiliated with respective schools of thought in economic history[10]:

(a) The *formalist* position analyzes communities by applying the neoclassical model of utility maximization under scarcity. The approach is remarkably similar to cliometrics.
(b) The *substantivist* position was developed by Karl Polanyi, according to whom economic analysis of non-western societies should be carried out with respect to that society's history, within its own socio-cultural and political context. The influence of the German School is obvious. The economy is considered "inferior" to other institutions and traditions and social research should therefore focus primarily on studying the latter. The substantivist approach has many similarities to the neo-Marxist theories of dependence which heavily influenced the economic thought in Latin America.
(c) *Culturalism* is a version of the substantivist position, claiming that each community should be understood as a distinctive formation from both a historical and a cultural point of view. Any attempt to analyze and evaluate economic behaviors with the western criteria of rationalism and utility is bound to fail, since such criteria may be totally alien and incompatible with the norms adopted by the community under study. This approach is linked to "positive economics" as well, interested only in factual and descriptive accounts of events and avoiding *a priori* judgments based on optimization and utility.

15.3.6 Counterfactual History

This approach studies the causes of historical events and attempts to answer "*what if*" questions—what would have happened if crucial events had turned out differently. The lengths and credibility of counterfactuals determine the value of such comparative analysis and how fascinating virtual history can become.

Counterfactual stories often appear more like historical fiction rather than serious analytical efforts. Nevertheless, many distinguished historians, after analyzing the actual historical facts, legitimately change a number of crucial parameters to investigate alternative historical outcomes. A modern supporter of the alternative reading of history is Niall Ferguson[11] who dismisses both the teleological and the historicist positions and compares actual historical events with alternative ones. In his counterfactual accounts he argues that, contrary to

[10] For a detailed account of the various approaches, see Wilk and Cligget, *Economies and Cultures: Foundations of Economic Anthropology* (2007).
[11] Aptly titled *Virtual History* (2011).

the deterministic systems we find in nature, societies are more related to the dynamics of chaos, where even the slightest change of initial conditions can lead to strikingly different outcomes. To avoid any conceptual confusion with actual events, he coined the term "*chaostory*" to express the multitude of alternative outcomes. What has actually happened is just one of the paths that could have been followed.

Perhaps this approach is best suited to utilize the lessons that can be drawn from economic history and the history of economic ideas, and provide insights on how new theories are formulated.

References

Blaug M (1980) The methodology of economics: or how economists explain. Cambridge University Press, Cambridge

Boldizzoni F (2011) The poverty of Clio: resurrecting economic history. Princeton University Press, Princeton, NJ

Ferguson N (2011) Virtual history. Penguin, London

Friedman M (1979) The methodology of positive economics. In: Hahn F, Hollis M (eds) Philosophy and economic theory. Oxford University Press, Oxford

Marcuse H (1983) From Luther to Popper. Verso, London

Olson M (1984) The rise and decline of nations: economic growth, stagflation and social rigidities. Yale University Press, New Haven, CT

Popper K (1999) All life is problem solving. Routledge, New York

Reichenbach H (1970) From Copernicus to Einstein. Dover, New York

Wilk R, Cligget L (2007) Economies and cultures: foundations of economic anthropology. Westview Perseus, Boulder, CO

MIX
Papier aus verantwortungsvollen Quellen
Paper from responsible sources
FSC® C105338

If you have any concerns about our products,
you can contact us on
ProductSafety@springernature.com

In case Publisher is established outside the EU,
the EU authorized representative is:
Springer Nature Customer Service Center GmbH
Europaplatz 3, 69115 Heidelberg, Germany

Printed by Libri Plureos GmbH
in Hamburg, Germany